Mites of Economic Plants: Identification, Bio-ecology and Control

Mites of Economic Plants: Identification, Bio-ecology and Control

Editor

Pradnya Jain

scitus
academics

Mites of Economic Plants: Identification, Bio-ecology and Control

Edited by **Pradnya Jain**

ISBN: 978-1-68117-079-4

Library of Congress Control Number: 2015935477

SCITUS Academics LLC,
www.scitusacademics.com
Box No. 4766, 616 Corporate Way,
Suite 2, Valley Cottage,
NY 10989

Printed in United States of America on Acid Free Paper ∞

Contents

Preface

Mites are among the most important arthropods in greenhouses, both as pests causing economic injury to greenhouse crops and natural enemies used in the biological control of pest insects and mites in greenhouses. Because of their minute size, mites are much less well known than insects. Although a great deal has been written on mites of greenhouses in widely dispersed scientific literature, there is a lack of a comprehensive treatment dedicated to greenhouse mites. The need for such a book on the identification, biology and control of greenhouse mites has been brought to my attention by students, entomologists, pest control workers, and growers, indirectly through correspondence and directly through interactions with them in training courses and collaborative research. Control studies included the evaluation of field efficacies of seven pesticides against different stages. Mites pose a serious problem to plants worldwide, attacking crops and spreading disease. When mites damage crops of economic importance the impacts can be felt globally. Mites are among the most diverse and successful of invertebrates, with over 45,000 described species, with many more thousands to be discovered. They are responsible for a significant portion of the losses of crops for food, fiber, industry and other purposes, and require expensive and often controversial pest control measures. Understanding these mites is vital for entomologists, pest researchers, agronomists and food producers. Knowledge of mite pests helps to inform control strategies and optimize the production of economic plants and the agrarian economy. This book provides a thorough coverage of the mites and the problems they cause to crops, yet it is easily searchable, organized by mite species and subdivided into helpful headings. It takes a worldwide view of the issue of mites injurious to economic plants, describing mites prevalent in different regions and discussing control methods appropriate in different environments.

Editor

Chapter 1

Ecological and Economic Importance of Bats (Order Chiroptera)

Mohammed Kasso and Mundanthra Balakrishnan

Department of Zoological Sciences, Addis Ababa University, Addis Ababa, Ethiopia

ABSTRACT

Order Chiroptera is the second most diverse and abundant order of mammals with great physiological and ecological diversity. They play important ecological roles as prey and predator, arthropod suppression, seed dispersal, pollination, material and nutrient distribution, and recycle. They have great advantage and disadvantage in economic terms. The economic benefits obtained from bats include biological pest control, plant pollination, seed dispersal, guano mining, bush meat and medicine, aesthetic and bat watching tourism, and education and research. Even though bats are among gentle animals providing many positive ecological and economic benefits, few species have negative effects. They cause damage on human, livestock, agricultural crops, building, and infrastructure. They also cause airplane strike, disease transmission, and contamination, and bite humans during self-defense. Bat populations appear to be declining presumably in response to human induced environmental stresses like habitat destruction and fragmentation, disturbance to caves, depletion of food resources, overhunting for bush meat and persecution, increased use of pesticides, infectious disease, and wind energy turbine. As bats are among the most overlooked in spite of their economical and ecological importance, their conservation is mandatory.

INTRODUCTION

The order Chiroptera is the second most diverse among mammalian orders, which exhibits great physiological and ecological diversity [1]. They form one of the largest nonhuman aggregations and the most abundant groups of mammals when measured in numbers of individuals [2]. They evolved before 52 million years ago and diversified into more than 1,232 extant species [3]. They are small, with adult masses ranging from 2 g to 1 kg; although most living bats weigh less than 50g as adults [4]. They have evolved into an incredibly rich diversity of roosting and feeding habits. Many species of bats roost during the day time in foliage, caves, rock crevices, hollows of trees, beneath exfoliating bark, and different man-made structures [2]. During night, they become active and forage on diverse food items like insects, nectar, fruits, seeds, frogs, fish, small mammals, and even blood [3].

The forelimb of a bat is modified into a wing with elongated finger bones joined together by a thin and large (85% of the total body surface area) membrane with rich blood flow [5]. Their wing is an unusual structure in mammals enabling for active unique powered flight. Skin covering the wings of bats not only constitutes a load-bearing area that enables flying but also performs multiple functions like providing a protective barrier against microbes and parasites, gas exchange, thermoregulation, water control, trapping of insects, and food manipulation and for swimming [6]. The powerful flight of bats plays the most important role for their widespread distribution and diversity. This helps in the occurrence of bats in all continents except

Antarctica, some Polar Regions, and some isolated oceanic islands. It has also contributed a lot for their extraordinary feeding and roosting habits, reproductive strategies, and social behaviors [2].

Although all bats do not echolocate, in general echolocation is considered as one of the major characteristics of bats. Even if the role of echolocation for plant-visiting bats is not clear, they use wide range (10–200 kHz) of ultrasonic frequencies during foraging. The availability of commercially produced bat detectors contributed a lot in linking data of echolocation with the biology of bats [3].

Bats are an essential natural resource that play great role in providing many ecological and economic services [7]. However, the determination of the ecological and economic values provided by bats is extremely challenging except from the studies on ecosystem services provided directly to the production of goods and services consumed by humans [3, 7].

ECOLOGICAL IMPORTANCE OF BATS

Bats have long been postulated to play important ecological roles in prey and predator, arthropod suppression, seed dispersal, pollination, material and nutrient distribution, and recycle [3].

As Predators

Bats have diverse patterns of feeding in which some select among available prey while others are generalist predators, feeding on a wide diversity of taxonomic groups. They also opportunistically consume appropriately sized prey depending on availability within a preferred habitat [8]. Their prey size can vary from 1 mm (midges and mosquitoes) to as large as 50 mm long (beetles and large moths) based on the species of bat [8, 9].

Remains of 12 orders or classes of prey belonging to 18 taxonomic families of insects were reported in the diet of bats [10]. The prey items include Acari, Arachnida, Coleoptera, Diptera, Hemiptera, Homoptera, Hymenoptera, Isoptera, Lepidoptera, Neuroptera, Orthoptera, and

Trichoptera. They also predate on frogs, fish, small mammals, and even blood of mammals and birds [3]. Some species also eat unusual prey items such as scorpions and spiders [11]. Bats exhibit high species diversity with multiple species forage sympatrically to avoid competition. A resource partition is possible through the use of diverse mechanism like difference in wing shape, body size, and sensory cues [12].

Obtaining accurate estimates of the amount of prey consumed by bats is challenging. However, its amount and type are confirmed as it varies with prey availability, time during night, species, sex, age, and the reproductive status of bats [13, 14]. Variety of approaches like direct observation [15], comparison of pre- and postflight body mass [14], and fecal sample analysis [16] have been used to estimate the amount and type of prey consumed by bats. Results of studies carried out on insectivorous bats indicated that they consume more than 25% of their body mass of insects each night [17]. At the peak night of lactation, a 7.9 g little brown bat (Myotis lucifugus) needs to consume 9.9 g of insects which is over 100% of its body mass [18]. At peak lactation, a female Brazilian free-tailed bat (Tadarida brasiliensis) consumes insects up to 70% of the body mass each night. It frequently selects nutrient-rich abdomen of moths while discarding the wings, head, and appendages, which greatly increases feeding efficiency and the quantity of insects consumed [14]. This can indicate that maternity colony of one million Brazilian free-tailed bats weighing 12 g each could prey up to 8.4 metric tons of insects in a single night. These studies hint at the immense capability of insect consumption and the potential role of bats in the suppression of arthropod populations [3]. Based on fecal sample analyses, a colony of 300 evening bats (Nycticeius humeralis) and 150 big brown bats in Indiana was estimated to consume 6.3 and 1.3 million insects per year, respectively [16, 19].

In this way, an estimated 99% of potential crop pests are limited by natural ecosystems of which some fraction can be attributed to predation by bats [7]. Predation of bats can have direct effects on herbivore communities and indirect effects on plant communities through both density mediated (consumption) and trait-mediated (behavioral) interactions and for nature balance [20].

Prey for Vertebrates

Although there are relatively few observations of animals feeding on bats, a number of vertebrate predators like fish, amphibians, birds, reptiles, and mammals prey on bats throughout the world [21, 22]. The main bat predators are owls, hawks, falcons, snakes, and mammals such as raccoons, ringtails, and opossums. In some countries like New Zealand, forest-floor dweller bats are frequently predated by the introduced rats, feral cats, and weasel [23]. The larger phyllostomid bats (Vampyrum spectrum, Chrotopterus auritus, and Phyllostomus hastatus) are known to eat smaller bats [24].

Bats generally comprise a relatively small proportion of the diet of most predators. Bats represented only 0.003% of the diet of small falcons and hawks and 0.036% of the diet of owls in Great Britain [21]. Although diurnal raptors feed on bats during twilight hours in some parts of the world [25], nocturnal predation by owls is the most significant predation pressure on bats in temperate regions [21].

Most of the bats are predated on roosting or when they emerge from roosts although sometimes predated during foraging or flying. Large concentrations of bats at roost sites and the relatively predictable patterns of their emergence from roosts, provide significant opportunities for predators to prey on bats [25]. However, strategies like low dependability to roost sites, selection of time, and patterns of emergence from roosts and nocturnal activity are used to minimize the risk of predation [21].

Hosts for Parasite

Numerous haematophagous ectoparasites live such as bat fleas (Ischnopsyllidae), bat flies (Nycteribiidae), bat mites (Spinturnicidae), and bugs (Cimicidae) on the skin surface and in the fur of bats. These obligate ectoparasites are specialized to their hosts [26]. The skin and hair morphology play important roles in affecting the parasite's life style in terms of adaptation, feeding, movement and egg laying resulted in morphological adaptations with coevolution of both species [5].

The hair density as well as surface structures of bat hairs and the distribution of mast cells are very important for the host defense against parasite infestation. Although the hair density of bats primarily provides protection against unfavorable microclimatic conditions, it also serves as passive antiparasitic defense. The high hair density prohibits infestation by large parasites. However, dense fur in some parts of the host's body may provide a suitable shelter for specialized small parasites [5].

Pollination

In addition to insect suppression through predation, some bat species primarily the two families of bats (Pteropodidae in the Old World and Phyllostomidae in the New World) play important roles in plant pollination [3]. Although bat pollination is relatively uncommon when compared with bird or insect pollination, it involves an impressive number of economically and ecologically important plants [27]. Particularly, beyond the economic value of plant pollination and seed dispersal services, plant-visiting bats provide important ecological services by facilitating the reproductive success and the recruitment of new seedlings [3]. Many of these plants are among the most important species in terms of biomass in their habitats. For instance, bat-pollinated columnar cacti and agaves are dominant vegetation elements in arid and semiarid habitats of the New World [3].

Bat pollination occurs in more than 528 species of 67 families and 28 orders of angiosperms worldwide [28]. Pteropodid bats are known to pollinate flowers of about 168 species of 100 genera and 41 families and phyllostomid bats pollinate flowers of about 360 species of 159 genera and 44 families [28]. As feeding on nectar and pollen requires relatively specialized morphology (e.g., elongated snout and tongue), relatively few members of these families are obligate pollinators. Unlike predation, which is an antagonistic population interaction, pollination, and seed dispersal are mutualistic population interactions in which plants provide a nutritional reward (nectar, pollen, and fruit pulp) for a beneficial service [3].

Seed Dispersal

Seed dispersal is a major way in which animals contribute for ecosystem succession by depositing seeds from one area to another [29]. As 50–90% of tropical trees and shrubs produce fleshy fruits adapted for consumption by vertebrates, the role played by frugivorous bats in dispersing these seeds is tremendous [30].

Countless tropical trees and understory shrubs are adapted for seed dispersal by animals, primarily by bats and birds. Particularly, night-foraging fruit bats are more compliant than birds by covering long distances each night, defecating in flight, and scattering far more seeds across cleared areas [31]. Unlike most seed dispersal by vertebrates that dispersed close to parent plants with only 100–1,000 m away, the seeds dispersed by frugivorous bats were relatively far away (1-2 km) [31]. Furthermore, the flying fox migration for more than 1,000 km across the central belt of the African continent helps to scatter huge numbers of seeds along the way. Unlike birds, bats tend to defecate or spit out seeds during flight and hence facilitate seed dispersal in clear-cut strips [32]. In addition to their tendency to defecate seeds in flight, many bats use one or more feeding roosts each night where they deposit the vast majority of seeds ingested far away from fruiting plants.

Many bat-dispersed seeds are from hardy pioneer plants, the first to grow in the hot, dry conditions of clearings with up to 95% chance of germination. As these plants grow, they provide shelter that helps other, more delicate plants to grow [33]. Fruit-eating bats play an extremely important role in forest regeneration. Tropical frugivorous bats also facilitate tropical forest regeneration and help to maintain species diversity by introducing seeds from outside disturbed areas, whereas the neotropics frugivorous bats play important role in the early stages of forest succession [32].

The dispersed seeds of palms and figs by bats are also common in many tropical forests. Because they are also eaten by many birds and mammals, figs often act as keystone species in tropical forests [34].

Soil Fertility and Nutrient Distribution

Bats play an important ecological role in soil fertility and nutrient distribution due to their relatively high mobility and the use of different habitats for roosting and foraging, which facilitates nutrient transfer within ecosystems [35, 36]. However, the suspected importance of nutrient transfer by bats in overall ecosystem function is probably relatively low when compared with microhabitat conditions [36]. For soil fertility and nutrient distribution, bat guano has a great ecological potential as bats sprinkle it over the landscape throughout the night. Thus, bats contribute a lot in nutrient redistribution, from nutrient-rich sources (e.g., lakes and rivers) to nutrient-poor regions (e.g., arid or upland landscapes) [35]. For instance, a colony of one million Brazilian free-tailed bats (T. brasiliensis) in Texas can contribute to 22 kg of nitrogen in the form of guano.

Bat guano in turn supports a great diversity of organisms including arthropods, fungi, bacteria, and lichens that represent different trophic levels [37]. The diversity of organisms living on guano differs depending on the species and their diet. For example, guano from insectivorous bats is typically inhabited by mites, pseudoscorpions, beetles, thrips, moths, and flies, whereas the guano of frugivorous bats is inhabited by spiders, mites, isopods, millipedes, centipedes, springtails, barklice, true bugs, and beetles [38]. As bats regularly or occasionally roost in caves, bat guano provides the primary organic input to cave ecosystems, which are inherently devoid of primary productivity. They provide essential organic input that supports assemblages of different endemic cave flora and fauna. For example, cave-dwelling salamander and fish populations and invertebrate communities are also highly dependent upon nutrients from bat guano. However, little consideration has been given to the role of bats in supporting entire cave ecosystems [39].

Bioindicators

The earth is now subject to climate change and habitat deterioration on a large scale. Monitoring of climate change and habitat loss alone is insufficient to understand the effects of these factors on complex biological communities [40]. Ecosystems are geographically variable and inherently complex whereas responses to anthropogenic changes are in a nonlinear and scale dependent manner. Thus, a broad-scale network of monitoring that captures local, regional, and global components of the earth's biota is critical for understanding and forecasting responses to climate change and habitat conversion [2]. It is therefore important to identify bioindicator taxa that show measurable responses to climate change and habitat loss and that reflect wider-scale impacts on biodiversity [2].

There are three types of bioindicators (biodiversity, ecological, and environmental indicators) [41]. Biodiversity indicators capture responses of a range of taxa and reflect components of biological diversity such as species richness and species diversity. Ecological indicators consist of taxa or assemblages that are sensitive to identified environmental stress factors that demonstrate the effect of those stress factors on biota. Environmental indicators respond in predictable ways to specific environmental disturbances [41].

Biodiversity indicator species have characteristics that can be used as an index of attributes (e.g., presence/absence, population density, and relative abundance) of other species comprising the biota of interest [42]. Thus, these species collectively must have characteristics that make them easily identifiable (stable taxonomy), easy to sample, and show graded responses to habitat degradation that correlate with the responses of other taxa [43]. In addition, as environmental degradation can occur over a variety of scales, monitoring the impacts of such threats through indicator species requires the species that have broad geographic ranges. Bats, as volant taxa, fulfill this criterion better than most other taxa [2].

Bats are excellent ecological indicators of habitat quality. They have enormous potential as bioindicators to both disturbance and the existence of contaminants due to a combination of their size, mobility, longevity, taxonomic stability, observable short and long term effects, trends of populations, and their distribution around the globe [2, 4, 44].

Bat populations are affected by a wide range of stressors that affect many other taxa. In particular, changes in bat numbers or activity can be related to climate change (including extremes of drought, heat, cold, precipitation, cyclone, and sea level rise), deterioration of water quality, agricultural intensification, loss and fragmentation of habitats, fatalities at wind turbines, disease, pesticide use, and overhunting [2]. The magnitude of changes around the globe is quite variable as is the nature of the human activities that alter and fragment landscapes differs from one place to another [45]. As insectivorous bats occupy high trophic levels, they are sensitive to accumulations of pesticides and other toxins, and changes in their abundance may reflect changes in populations of arthropod prey species [1]. High fatalities observed in bats associated with diseases, may provide an early warning of environmental links among contamination, disease prevalence, and mortality. Increased environmental stress can suppress the immune systems of bats and other animals and thus one might predict that the increased prevalence of diseases is a consequence of altered environments [2].

Economic Importance of Bats

Biological Pest Control

Among the estimated 1,232 extant bat species, over two-thirds are either obligate or facultative insectivorous mammals. They consume nocturnal and crepuscular species of insects from different habitats as such forests, grasslands, agricultural landscapes, aquatic, and wetland habitats [3].

Various species of prominent insect pests have been found in the diet of bats based on identification of insect fragments in fecal samples and stomach contents. They consume enormous quantities of insect pests that cost farmers and foresters billions of dollars annually [46]. These insects include, June beetles (Scarabidae), click beetles (Elateridae), leafhoppers (Cicadelidae), plant hoppers (Delphacidae), the spotted cucumber beetle (Chrysomelidae), the Asiatic oak weevil (Curculionidae), and the green stinkbug (Pentatomidae) [3].

Mexican free-tailed bats (T. brasiliensis) feed an estimated one million kilogram of the most costly agricultural pest insects (corn earworm moth) each night [47]. One bat can eat 20 female corn earworm moths in a night and each moth can lay as many as 500 eggs, potentially producing 10,000 crop-damaging caterpillars [46]. About 150 big brown bats also consume enough adult cucumber beetles in one summer to prevent egg-laying that could produce 33 million root-worm larvae and contributing in prevention of agricultural pests damage [16]. Thus, the death of one million bats from the disease called white nose syndrome indicates 660–1,320 metric tons of insects are no longer being consumed each year in affected areas [36]. Millions of Brazilian free-tailed bats each evening consume a wide variety of prey items (12 orders, 35 families) of about 14,000 kg agricultural pests [48, 49]. Based on the dietary composition (minimum number of the total insects per guano pellet), number of specific agricultural pest species in each pellet, and the number of active foraging days per year, a colony of 150 big brown bats (Eptesicus fuscus) in the midwestern United States annually consume approximately 600,000 cucumber beetles, 194,000 June beetles, 158,000 leafhoppers, and 335,000 stinkbugs, which are severe crop pests [16].

Bats are just one of several groups of animals that naturally prey upon mosquitoes. A Florida colony of 30,000 southeastern myotis (Myotis austroriparius) eats 50 tons of insects annually, including more than 15 tons of mosquitoes [8]. It is also known that northern long-eared bats (Myotis septentrionalis) suppress mosquito populations through direct predation [50].

The estimation of the economic importance of bats in agricultural systems is challenging [36]. A common challenge in the study of the use of bats as pest control is the lack of basic ecological information regarding foraging behavior and diet for many species of bats. For example, traditional dietary analyses through fecal or stomach contents have only identified arthropod fragments to the ordinal or familial level, rather than to species [9, 17] and in cases where species identification is possible, it has typically been restricted to hard-bodied insects although recent novel molecular techniques have allowed detection and species identification of both hard and soft bodied insects [51, 52]. However, the value of pest suppression services provided by bats ranges from \$12 to \$173 per 0.405 ha in Texas [48]. In USA, the estimate value of bats as a result of reduced costs of pesticide applications due to insect pest suppression by bat predation is in the range of \$3.7–\$53 billion per year excluding the costs of impacts of pesticides on ecosystems [36].

Pollination

As pollinators, tropical bats provide invaluable support to many local and national economies [33]. Large-scale cash crops that are originally pollinated or dispersed by bats include wild bananas, mangos, breadfruits, agave, durians, and petai of which durians and petai currently rely on bats for pollination [7]. Durian, a wildly popular fruit worth more than \$230 million per year in southeast Asia, opens its flower at dusk and relies almost exclusively on fruit bats for pollination [7].

Except the "ornamental" bananas with upright flowers that are pollinated by birds, all the rest, including the ancestors of edible bananas, that have horizontal or drooping flowers are pollinated primarily by bats [33]. Their adaptations for bat pollination include nocturnal flowering, a strong and characteristic odor that attracts bats, plus abundant and accessible nectar and pollen. The coevolution of bananas and bats over 50 million years also resulted in adaptations for effective seed dispersal even if other mammals like monkeys feed on fruits and disperse seeds [33]. Although bats are no longer needed to pollinate flowers or disperse the seeds of edible bananas, the ecological services bats provide for their wild relatives are important for conserving its genetic diversity [3].

Agave macroacantha is extremely dependent on nocturnal pollinators for its reproductive success of which bats are especially important for its successful pollination [53]. Some of these pollinators (bats) are migratory, and have been reported to be steadily declining. A continuing decline in the populations of pollinators may hamper the successful sexual reproduction of the plant host and may put its survival under risk [53].

The Mahwa tree or honey tree (Madhuca indica) is pollinated by bats. These pollination services highlight one of the highly valued ecosystem services provided by plant-visiting bats both culturally and economically. The timber of this tree is used for making farm cart wheels in India. The flowers are used as food and for preparing a distilled spirit and its sun-dried fruits for human consumption and the oil extracted from flowers and seeds as ingredients for soaps, candles, cosmetics, lubricants, and medicines [54].

Similarly, there are 289 Old World tropical plant species that rely on pollination and seed dispersal services by bats for their propagation [7]. These plants, in turn, contribute to the production of 448 bat-dependent products in a variety of categories such as timber and other wood products (23%); food, drinks and fresh fruit (19%); medicine (15%); dye, fiber, animal fodder, fuel wood, ornamental plants, and others (43%). However, because bat-provided services represent one input within a multi-input production process, only a portion of the total value of the end product can be attributed to bats [7].

The pollination services of bats for 100 food crops by combining the pollination dependence ratios with regional crop production and its prices was determined [55]. Of these, 46 crops depended to some degree on animal pollinators (6 essentially dependent, 13 highly dependent, 13 moderately dependent, and 14 slightly dependent) accounting for 39% of world production value.

Based on the crop production and animal-dependent pollination, the total economic value of bats in global pollination services is estimated to be $200 billion, representing 9.5% of the value of world food crop production in 2005 [55].

Seed Dispersal

Bats are crucial to the survival of the world's tropical forests. Enormous expanses of rain forest are cleared every year for logging, agriculture, ranching, and other uses. Fruit-eating bats are uniquely suited for dispersing the seeds of "pioneer plants" from which a diverse and healthy forest can reemerge [33]. Thus, the economic value contributed by bats in maintaining forests is tremendous. For instance, the economic value estimate for seed dispersal services provided by bats to the regeneration of giant oak is $212,000 for seeding acorns and $945,000 for planting saplings [56]. The tropical almond tree, Terminalia catappa, is one of the bat-dispersed trees with many human uses like shade, fuel-wood, edible nuts, timber, and tannin (extracted from the bark, leaves, roots, and the fruit shell). The large leaves are also used as wrapping material and have also many medicinal uses, including diaphoretic, anti-indigestion, antidysentery and headache [33].

Guano Mining

Guano from bats has long been mined from caves for use as fertilizer on agricultural crops due to its high concentrations of limiting nutrients like nitrogen and phosphorous [57]. It provides some of the world's finest natural fertilizers [58]. About 950 bat guano products show a market demand for the product. Prices for bat guano organic fertilizer varied from $1.25 to $12.00 per 0.5 kg depending on the size of the package (larger packages have lower unit prices) and the mix of its ingredients [3]. The Mexican free-tailed bat guano has been extracted for fertilizer in thousands of tons from Bracken Cave in Texas alone with the current retail sales ranging from $2.86 to $12.10 per kilogram [58]. In some places, guano harvesting is carried out on a sustainable basis, especially in caves where bats normally migrate elsewhere for a part of their life each year. The bacteria extracted from bat guano have also been used by some companies to improve detergents and other products of great value to humans [58].

Bush Meat and Medicine

Bats have also long been used for food and medicine [59]. They provide a direct source of human food in many countries [60]. Several anecdotal price information of bat bush meat ranges from $2.50 to $3.50 per bat in Malaysia and $0.43–$10 per bat in Jakarta. Bat bush meat has the highest nutrient (high protein, vitamin and mineral composition) with lowest cost per kg [7, 61, 62]. Several studies have reported on the overhunting of bat for bush meat indicating a need for further conservation [61].

The anticoagulant compound called salivary plasminogen activator (DSPA) found in the saliva of the common vampire bat is used to treat strokes. Unlike alternative medicines, it can be administered even much later after a stroke has occurred and still be effective [63]. Physicians used bats to treat ailments of patients ranging from baldness to paralysis [60, 63].

Aesthetic and Bat Watching Tourism

Wildlife watching is simply an activity that involves watching wildlife to identify and observe their behavior and appreciate their beauty. It differs from other forms of wildlife-based activities like hunting and fishing [49]. Although perhaps not as widely practiced as bird watching, bat watching is currently growing as a recreational activity [49]. Similar to other wildlife watching tourism, it also generates income in the form of entrance and permit fees, personal payments to the guides, drivers and scouts and payment for accommodation, and other services [49].

The majority of bat watching takes place at cave entrances where bats emergence can be viewed. For this purpose, the charge ranges from $5 to $12 per visitor. For instance, the Congress Avenue Bridge, which is the home to the largest urban bat colony of approximately 1.5

million Mexican free-tailed bats (T. brasiliensis) in USA is visited by 200–1,500 visitors per evening with the value of $3 million per year. The spectacular flock emergence of bats from their roost from March to November, to feed and migrate south during the winter months serves as tourist attraction [49, 64].

Education and Research

Although extremely difficult to quantify, it is important to recognize the extraordinary value of bats to ancient and contemporary traditions and science. The current study of bat echolocation and locomotion has provided inspiration for novel technological advances in biomedical ultrasound, sensors for autonomous systems, and wireless communication and BATMAVs (bat-like motorized aerial vehicles) [65, 66].

Bats contributed a lot to the field of biomimetics, which is the science of modeling cutting-edge technologies based on natural forms [65]. The anticlotting chemicals in the saliva of bats are also currently being investigated as potential anticoagulant for people who are at high risk of blood clots and strokes. In addition, the development of sonar for ships and ultrasound was partly inspired from echolocation that bats use as navigation system to find and follow their prey at night without crashing on trees, buildings, or other obstructions [65, 66].

Particularly, a unique feature of bats that provides potential for future application is their flying ability by their own power. The aerodynamic range of bats includes changing flight direction by turning 180 degrees within just three wing beats while flying at full tilt. They are such quick flyers because of the quickness of their wings that are structured to fold during flight, similar to the way that a human hand folds. Also, their wings are draped by stretchy skin and are powered by special muscles. Ongoing research about the structure of bat wings and the mechanics of bat flight may ultimately lead to the development of technologies that improve the maneuver ability of airplanes [22].

Bats as Pests

Although bats are grouped among the world's gentlest animals that provide many positive ecological and economic benefits, few of them are considered as pests. They may cause damage on human, livestock, agricultural crops, airplane strike, building, and infrastructure infestation, and rarely become aggressive or bite humans during self-defense [58]. For instance, frugivorous bats that feed on some economically important fruits result in greater loss.

Three species of vampire bats that occur in the New World are major pests feeding mainly on the blood of livestock (cattle, equines, goats, sheep, and pigs), poultry and occasionally humans. They are also responsible in transmitting rabies. Populations of vampire bats have increased sharply in areas of Latin America where European livestock have been introduced [67]. Wounds caused by vampire bats may also be vulnerable to secondary infections [1].

Bat strikes to airplane have been responsible for loss of human lives and damage to materials worldwide resulting in loss of billions of dollars annually [68, 69]. For example, 821 bat strikes were reported in the USA Air Force during 1997–2007 [70] and 327 from 91 airports during 1996–2006 in Australia [71]. From less than 1% of the bat-strike reported in USA, a cumulative damage is more than $825,000 of which more than half is attributable to 5 bat-strike incidents [70]. This high damage is accredited to high body weight (up to 1 kg like flying foxes) and unlike birds, they possess none pneumatized solid and heavy bones. It results in a greater and more concentrated impact force of strike and a greater capacity to perforate an aircraft than bird strikes [69]. Australian flying foxes roost gregariously and emerge from roosts in flocks, which may include thousands of flying foxes, thus the increasing risk of multiple simultaneous strikes. Their major damage to aircraft includes breaking of windscreen, perforation of aircraft skin, and ingestion into engine [71].

Building and house infestation by bat constitutes a serious public health problem [72]. They spoil food and make ceilings, walls, and floors dirty with the accumulation of guano and urine [72]. Besides, they cause discomfort to humans by their distressing noise, offensive odors, and attraction of coprophagous insects. Potential health hazards may result from chitinous remains of finely chewed insects in guano, attack of ectoparasites, drinking water contamination by urine and feces [72].

Disease Transition and Contamination

Bats are hosts to a range of zoonotic and potentially zoonotic pathogens. They differ from other disease reservoirs because of their unique and diverse lifestyles, including their ability to fly, often highly gregarious social structures, long life spans, and low fecundity rates [73]. They represent a potential epidemiologic of several diseases that can be fatal to humans, including rabies, Ebola, leptospirosis, histoplasmosis, and pseudotuberculosis [72, 74].

Bats are reservoirs of several pathogens, whose spread may be related to physiological stress associated with habitat loss or alteration [75]. The recent die-offs of bats presenting with white nose syndrome may relate to increased levels of environmental stress that render them to be susceptible to fungal infection and viral infections like Henipaviruses, European bat lyssaviruses, rabies, and Ebola virus [74].

Human activities that increase exposure to bats will likely increase the opportunity for infections [73]. Like bird droppings, bat guano can contain a potentially infectious fungus Histoplasma capsulatum that causes lung infection known as histoplasmosis [58].

Bat populations appear to be declining presumably in response to human induced environmental stresses like habitat destruction and fragmentation, disturbance to caves, depletion of food resources, overhunting for bush meat and persecution, increased use of pesticides, infectious disease, and wind energy turbine. As bats are among the most overlooked in spite of their economical and ecological importance, their conservation is mandatory.

ACKNOWLEDGMENTS

The authors are grateful to Professor Afework Bekele for his valuable comments, suggestions, and corrections on the draft of this paper.

REFERENCES

1. A. M. Hutson, S. P. Mickleburgh, and P. A. Racey, Microchiropteran Bats: Global Status Survey and Conservation Action Plan, IUCN/

SSC chiroptera specialist group, IUCN, Gland, Switzerland, 2001.

2. G. Jones, D. Jacobs, T. H. Kunz, M. R. Wilig, and P. A. Racey, "Carpe Noctem: the importance of bats as bioindicators," Endangered Species Research, vol. 8, pp. 3–115, 2009.
3. T. H. Kunz, E. B. de Torrez, D. Bauer, T. Lobova, and T. H. Fleming, "Ecosystem services provided by bats," Annals of the New York Academy of Sciences, vol. 1223, no. 1, pp. 1–38, 2011.
4. M. B. Fenton, "Science and the conservation of bats: where to next?" Wildlife Society Bulletin, vol. 31, no. 1, pp. 6–15, 2003.
5. J. P. Madej, L. Mikulova, A. Gorosova et al., "Skin structure and hair morphology of different body parts in the Common Pipistrelle (Pipistrellus pipistrellus)," Acta Zoologica, vol. 94, no. 4, pp. 478–489, 2012.
6. A. N. Makanya and J. P. Mortola, "The structural design of the bat wing web and its possible role in gas exchange," Journal of Anatomy, vol. 211, no. 6, pp. 687–697, 2007.
7. M. S. Fujita and M. D. Tuttle, "Flying foxes (Chiroptera: Pteropodidae): threatened animals of key ecological and economic importance," Conservation Biology, vol. 5, no. 4, pp. 455–463, 1991.
8. E. L. P. Anthony and T. H. Kunz, "Feeding strategies of the little brown bat (Myotis lucifugus) in southern New Hampshire," Ecology, vol. 58, pp. 775–786, 1977.
9. A. Kurta and J. O. Whitaker Jr., "Diet of the endangered Indiana bat (Myotis sodalis) on the northern edge of its range," The American Midland Naturalist, vol. 140, no. 2, pp. 280–286, 1998.
10. M. J. Lacki, J. S. Johnson, L. E. Dodd, and M. D. Baker, "Prey consumption of insectivorous bats in coniferous forests of north-central Idaho," Northwest Science, vol. 81, no. 3, pp. 199–205, 2007.
11. M. Holderied, C. Korine, and T. Moritz, "Hemprich's long-eared bat (Otonycteris hemprichii) as a predator of scorpions: whispering echolocation, passive gleaning and prey selection," Journal of Comparative Physiology A, vol. 197, no. 5, pp. 425–433, 2011.

12. D. Fukui, K. Okazaki, and K. Maeda, "Diet of three sympatric insectivorous bat species on Ishigaki Island, Japan," Endangered Species Research, vol. 8, no. 1-2, pp. 117–128, 2009.
13. T. H. Kunz, "Feeding ecology of a temperate insectivorous bat (Myotis velifer)," Ecology, vol. 55, pp. 693–711, 1974.
14. T. H. Kunz, J. O. Whitaker, and M. D. Wadanoli, "Dietary energetics of the insectivorous Mexican free-tailed bat (Tadarida brasiliensis) during pregnancy and lactation," Oecologia, vol. 101, no. 4, pp. 407–415, 1995.
15. M. B. C. Hickey and M. B. Fenton, "Behavioural and thermoregulatory responses of female hoary bats,Lasiurus cinereus (Chiroptera: Vespertilionidae), to variations in prey availability," Ecoscience, vol. 3, no. 4, pp. 414–422, 1996.
16. J. O. Whitaker, "Food of the big brown bat Eptesicus fuscus from maternity colonies in Indiana and Illinois," American Midland Naturalist, vol. 134, no. 2, pp. 346–360, 1995.
17. R. A. Coutts, M. B. Fenton, and E. Glen, "Food intake by captive Myotis lucifugus and Eptesicus fuscus(Chiroptera: Vespertilionidae)," Journal of Mammalogy, vol. 54, pp. 985–990, 1973.
18. A. Kurta, G. Bell, K. Nagy, and T. Kunz, "Energetics of pregnancy and lactation in free-ranging little brown bats (Myotis lucifugus)," Physiological Zoology, vol. 62, no. 3, pp. 804–818, 1989.
19. J. O. Whitaker and P. Clem, "Food of the evening bat Nycticeius humeralis from Indiana," The American Midland Naturalist, vol. 127, pp. 211–217, 1992.
20. O. J. Schmitz and K. B. Suttle, "Effects of top predator species on direct and indirect interactions in a food web," Ecology, vol. 82, no. 7, pp. 2072–2081, 2001.
21. J. R. Speakman, "The impact of predation by birds on bat populations in the British Isles," Mammal Review, vol. 21, no. 3, pp. 123–142, 1991.
22. M. B. Fenton, "Constraint and flexibility—bats as predators, bats as prey," Symposia of the Zoological Society of London, vol. 67, pp. 277–289, 1995.
23. M. J. Daniel and G. R. Williams, "A survey of the distribution, seasonal activity and roost sites of New Zealand bats," New Zealand Journal of Ecology, vol. 7, pp. 9–25, 1984.

24. G. G. Goodwin and A. M. Greenhall, "A review of the bats of Trinidad and Tobago," Bulletin of the American Museum of Natural History, vol. 122, pp. 191–301, 1961.
25. M. B. Fenton, I. L. Rautenbach, S. E. Smith, C. M. Swanepoel, J. Grosell, and J. van Jaarsveld, "Raptors and bats: threats and opportunities," Animal Behaviour, vol. 48, no. 1, pp. 9–18, 1994.
26. K. Dittmar, M. L. Porter, S. Murray, and M. F. Whiting, "Molecular phylogenetic analysis of nycteribiid and streblid bat flies (Diptera: Brachycera, Calyptratae): implications for host associations and phylogeographic origins," Molecular Phylogenetics and Evolution, vol. 38, no. 1, pp. 155–170, 2006.
27. T. H. Fleming and N. Muchhala, "Nectar-feeding bird and bat niches in two worlds: pantropical comparisons of vertebrate pollination systems," Journal of Biogeography, vol. 35, no. 5, pp. 764–780, 2008.
28. T. H. Fleming, C. Geiselman, and W. J. Kress, "The evolution of bat pollination: a phylogenetic perspective," Annals of Botany, vol. 104, no. 6, pp. 1017–1043, 2009.
29. R. S. Duncan and C. A. Chapman, "Seed dispersal and potential forest succession in abandoned agriculture in tropical Africa," Ecological Applications, vol. 9, no. 3, pp. 998–1008, 1999.
30. H. F. Howe and J. Smallwood, "Ecology of seed dispersal," Annual Review of Ecology and Systematics, vol. 13, pp. 201–228, 1982.
31. M. A. Horner, T. H. Fleming, and C. T. Sahley, "Foraging behaviour and energetics of a nectar-feeding bat, Leptonycteris curasoae (Chiroptera: Phyllostomidae)," Journal of Zoology, vol. 244, no. 4, pp. 575–586, 1998.
32. R. Muscarella and T. H. Fleming, "The role of frugivorous bats in tropical forest succession," Biological Reviews, vol. 82, no. 4, pp. 573–590, 2007.
33. I. W. Buddenhagen, "Bats and disappearing wild bananas: can bats keep commercial bananas on supermarket shelves?" Bats, vol. 26, pp. 1–6, 2008.
34. M. Shanahan, S. So, S. G. Compton, and R. Corlett, "Fig-eating by vertebrate frugivores: a global review," Biological Reviews of the Cambridge Philosophical Society, vol. 76, no. 4, pp. 529–572, 2001.

35. E. R. Buchler, "Food transit time in Myotis lucifugus (Chiroptera: Vespertilionidae)," Journal of Mammalogy, vol. 54, pp. 985–990, 1975.
36. J. G. Boyles, P. M. Cryan, G. F. McCracken, and T. H. Kunz, "Economic importance of bats in agriculture," Science, vol. 332, no. 6025, pp. 41–42, 2011.
37. G. A. Polis, W. B. Anderson, and R. D. Holt, "Toward an integration of landscape and food web ecology: the dynamics of spatially subsidized food webs," Annual Review of Ecology and Systematics, vol. 28, pp. 289–316, 1997.
38. R. L. Ferreira and R. P. Martins, "Diversity and distribution of spiders associated with bat guano piles in Morrinho cave (Bahia State, Brazil)," Diversity and Distributions, vol. 4, no. 5-6, pp. 235–241, 1998.
39. D. B. Fenolio, G. O. Graening, B. A. Collier, and J. F. Stout, "Coprophagy in a cave-adapted salamander; the importance of bat guano examined through nutritional and stable isotope analyses," Proceedings of the Royal Society B, vol. 273, no. 1585, pp. 439–443, 2006.
40. C. Parmesan, "Ecological and evolutionary responses to recent climate change," Annual Review of Ecology, Evolution, and Systematics, vol. 37, pp. 637–669, 2006.
41. M. A. McGeoch, "The selection, testing and application of terrestrial insects as bioindicators," Biological Reviews of the Cambridge Philosophical Society, vol. 73, no. 2, pp. 181–201, 1998.
42. P. B. Landres, J. Verner, and J. W. Thomas, "Ecological uses of vertebrate indicator species: a critique,"Conservation Biology, vol. 2, pp. 316–327, 1988.
43. C. E. Moreno, G. Sánchez-Rojas, E. Pineda, and F. Escobar, "Shortcuts for biodiversity evaluation: a review of terminology and recommendations for the use of target groups, bioindicators and surrogates,"International Journal of Environment and Health, vol. 1, no. 1, pp. 71–86, 2007.
44. M. B. Fenton, L. Acharya, D. Audet et al., "Phyllostomid bats (Chiroptera: Phyllostomidae) as indicators of habitat disruption in the neotropics," Biotropica, vol. 24, no. 3, pp. 440–446, 1992.

45. P. M. Vitousek, H. A. Mooney, J. Lubchenco, and J. M. Melillo, "Human domination of Earth's ecosystems," Science, vol. 277, no. 5325, pp. 494–499, 1997.
46. B. W. Keeley and M. D. Tuttle, Bats in American Bridges, vol. 4, Bat Conservation International, 1999.
47. G. F. McCracken, "Bats aloft: a study of high-altitude feeding," BATS, vol. 14, pp. 7–101, 1996.
48. C. J. Cleveland, M. Betke, P. Federico et al., "Economic value of the pest control service provided by Brazilian free-tailed bats in south-central Texas," Frontiers in Ecology and the Environment, vol. 4, no. 5, pp. 238–243, 2006.
49. R. Tapper, Wildlife Watching and Tourism: A Study on the Beneflts and Risks of a Fast Growing Tourism Activity and its Impacts on Species, UNEP/CMS Secretariat, Bonn, Germany, 2006.
50. M. H. Reiskind and M. A. Wund, "Experimental assessment of the impacts of northern long-eared bats on ovipositing Culex (Diptera: Culicidae) mosquitoes," Journal of Medical Entomology, vol. 46, no. 5, pp. 1037–1044, 2009.
51. G. F. McCracken, V. A. Brown, M. Eldridge, and J. K. Westbrook, "The use of fecal DNA to verify and quantify the consumption of agricultural pests," Bat Research News, vol. 46, pp. 195–196, 2005.
52. E. L. Clare, E. E. Fraser, H. E. Braid, M. H. Fenton, and P. D. Hebert, "Species on the menu of a generalist predator, the eastern red bat (Lasiurus borealis): using a molecular approach to detect arthropod prey," Molecular Ecology, vol. 18, no. 11, pp. 2532–2542, 2009.
53. S. Arizaga, E. Ezcurra, E. Peters, F. R. de Arellano, and E. Vega, "Pollination ecology of Agave macroacantha (Agavaceae) in a Mexican Tropical Desert: the role of pollinators," The American Journal of Botany, vol. 87, no. 7, pp. 1011–1017, 2000.
54. S. K. Godwa, R. C. Katiyar, and V. R. B. Sasfry, "Feeding value of Mahua (Madhuca indica) seed cakes in farm animals," Indian Journal of Dairy Science, vol. 49, pp. 143–154, 1996.
55. N. Gallai, J.-M. Salles, J. Settele, and B. E. Vaissière, "Economic valuation of the vulnerability of world agriculture confronted with pollinator decline," Ecological Economics, vol. 68, no. 3, pp. 810–821, 2009.

56. C. Hougner, J. Colding, and T. Söderqvist, "Economic valuation of a seed dispersal service in the Stockholm National Urban Park, Sweden," Ecological Economics, vol. 59, no. 3, pp. 364–374, 2006.
57. G. E. Hutchinson, "Survey of existing knowledge of biogeochemistry: the biogeochemistry of vertebrate excretion," Bulletin of the American Museum of Natural History, vol. 96, pp. 1–554, 1950.
58. M. D. Tuttle and A. Moreno, Cave-Dwelling Bats of Northern Mexico: Their Value and Conservation Needs, Bat Conservation International, Austin, Tex, USA, 2005.
59. T. P. Eiting and G. F. Gunnell, "Global completeness of the bat fossil record," Journal of Mammalian Evolution, vol. 16, no. 3, pp. 151–173, 2009.
60. S. Mickleburgh, K. Waylen, and P. Racey, "Bats as bushmeat: a global review," Oryx, vol. 43, no. 2, pp. 217–234, 2009.
61. F. O. Abulude, "Determination of the chemical composition of bush meats found in Nigeria," The American Journal of Food Technology, vol. 2, no. 3, pp. 153–160, 2007.
62. R. K. B. Jenkins and P. A. Racey, "Bats as bushmeat in Madagascar," Madagascar Wildlife Conservation, vol. 3, pp. 22–30, 2008.
63. W. D. Schleuning, "Vampire bat plasminogen activator DSPA-alpha-1 (desmoteplase): a thrombolytic drug optimized by natural selection," Pathophysiology of Haemostasis and Thrombosis, vol. 31, pp. 118–122, 2000.
64. L. A. Pennisi, S. M. Holland, and T. V. Stein, "Achieving bat conservation through tourism," Journal of Ecotourism, vol. 3, no. 3, pp. 195–207, 2004.
65. R. Müller and R. Kuc, "Biosonar-inspired technology: goals, challenges and insights," Bioinspiration and Biomimetics, vol. 2, pp. 146–161, 2007.
66. G. Bunget and S. Seelecke, "BATMAV: a 2-DOF bio-inspired flapping flight platform," in The International Society for Optics and Photonics, vol. 7643 of Proceedings of SPIE, pp. 1–11, 2010.
67. H. A. Delpietro, N. Marchevsky, and E. Simonetti, "Relative population densities and predation of the common vampire bat (Desmodus rotundus) in natural and cattle-raising areas in north-

east Argentina,"Preventive Veterinary Medicine, vol. 14, no. 1-2, pp. 13–20, 1992.

68. K. M. Brown, R. M. Erwin, M. E. Richmond, P. A. Buckley, J. T. Tanacredi, and D. Avrin, "Managing birds and controlling aircraft in the Kennedy Airport-Jamaica Bay Wildlife Refuge complex: the need for hard data and soft opinions," Environmental Management, vol. 28, no. 2, pp. 207–224, 2001.
69. Transport Canada, Sharing the Skies: An Aviation Industry Guide to the Management of Wildlife Hazards, Transport Canada, Ottawa, Canada, 2001.
70. S. C. Peurach, C. J. Dove, and L. Stepko, "A decade of U.S. Air Force bat strikes," Humboldt Wild life Care Center, vol. 3, pp. 199–207, 2009.
71. J. G. Parsons, D. Blair, J. Luly, and S. K. A. Robson, "Bat strikes in the Australian aviation industry,"Journal of Wildlife Management, vol. 73, no. 4, pp. 526–529, 2009.
72. A. M. Greenhall, "Bats: their public health importance and control with special reference to Trinidad," in Proceedings of the 2nd Vertebrate Pest Control Conference, vol. 18, pp. 108–116, 1964.
73. C. H. Calisher, J. E. Childs, H. E. Field, K. V. Holmes, and T. Schountz, "Bats: important reservoir hosts of emerging viruses," Clinical Microbiology Reviews, vol. 19, no. 3, pp. 531–545, 2006.
74. G. Wibbelt, A. Kurth, N. Yasmum et al., "Discovery of herpesviruses in bats," Journal of General Virology, vol. 88, no. 10, pp. 2651–2655, 2007.
75. M. B. Fenton, M. Davison, T. H. Kunz, G. F. McCracken, P. A. Racey, and M. D. Tuttle, "Linking bats to emerging diseases," Science, vol. 311, no. 5764, pp. 1098–1099, 2006.

Chapter 2

Evolutionary Aspects of Acaricide-Resistance Development in Spider Mites

Masahiro (Mh.) Osakabe1, Ryuji Uesugi1, and Koichi Goka2

[1]Laboratory of Ecological Information, Graduate School of Agriculture, Kyoto University, Kyoto 606-8502, Japan

[2]Invasive Alien Species Research Team, National Institute for Environmental Studies, Tsukuba 305-8506, Japan

ABSTRACT

Although the development of acaricide resistance in spider mites is a long-standing issue in agricultural fields, recent problems with acaricide resistance may be characterized by the development of complex- and/or multiresistance to acaricides in distinct classes. Such complexity of resistance is not likely to be a single mechanism. Pesticide resistance involves the microevolution of arthropod pests, and population genetics underlies the evolution. In this review, we address the genetic mechanisms of acaricide resistance evolution. We discuss genetic diversity and linkage of resistance genes, relationships between mite habitat and dispersal, and the effect of dispersal on population genetic structure and the dynamics of resistance genes. Finally, we attempt to present a comprehensive view of acaricide resistance evolution and suggest risks under globalization as well as possible approaches to managing acaricide resistance evolution or emergence.

INTRODUCTION

Safe and sufficient agricultural production is one of the most important global issues of the 21st century. Along the demands of food safety and minimizing environmental impacts, the idea of pest management is rapidly moving from chemical control toward sustainable technologies involving the exploitation of ecological functions. Nevertheless, insect pests and diseases cause huge yield losses in crops production, and pesticides have played an important role in mitigating yield loss of agricultural produce. Considering that the occurrence and geographical distribution of pests may change with global warming and economic globalization, pesticides will continue to be important in future agricultural production.

However, repeated pesticide use results in microevolution due to the selection pressure ultimately leading to resistance development. Marked development of acaricide resistance has been found in spider mites (Acari: Tetranychidae), which have short generation and high propagation potential. Since the 1990s, many populations of the two-spotted spider mite, Tetranychus urticae Koch, and the European red mite, Panonychus ulmi (Koch), have developed global resistance against many new acaricides. Among the top 20 resistant agricultural and medical arthropod pests, ranked by number of unique compounds in Arthropod Pesticide Resistance Database (http://www.pesticideresistance.org/), the most resistance has developed in T. urticae, whereas P. ulmi is the ninth most resistant [1].

Recent acaricide resistance problems are characterized by the development of complex- and/or multiresistance to acaricides in distinct classes. Pleiotropic effects may explain multiresistance in that a single key factor, for example, trans-acting factors, confers resistance to multiple insecticides involving distinct classes via a simultaneous increase in the action of several metabolic mechanisms involved in resistance [2]. Similarly, acaricide selection frequently confers cross resistance to acaricides in the distinct classes on spider mites, for example, T. urticae [3, 4] and P. ulmi [5]. However, the mechanisms of such resistance are largely unknown.

The complexity of acaricide resistance found today may not be explained by a single mechanism. Tetranychus urticae is a ubiquitous species that vigorously attacks an enormous number of plant species. Cross-resistance and the inheritance of resistance are not necessarily common to localities in this mite. These findings suggest that each selection process possibly leads to a different type of resistance mechanism, depending on the location and past selection history, and may yield different cross-resistance patterns.

Pesticide-resistance development is an example of microevolution in arthropod pests. Population genetics, which is affected by many factors including biological and ecological traits of arthropod pests, properties of pesticides, and pesticide application patterns [6–8], underlies the evolution. In this review, we address genetic mechanisms during the process of acaricide resistance evolution and discuss features of resistance genes and their genetic linkage, relationships between habitat and mite dispersal, and the effect of dispersal on population genetic structure and the dynamics of resistance genes to gain insight into spider mite resistance management in the global physical distribution environment. Finally, we comprehensively evaluate the evolution of acaricide resistance and suggest the global risks and management prospects.

ASPECTS AND FEATURES OF ACARICIDE RESISTANCE

Today's acaricide resistances assume a new complicated aspect rather than the past; that is, genetic mode of resistance is regionally different. Etoxazole resistance is inherited maternally in Korean T. urticae populations [9], implying that it is caused by a mutation in cytoplasmic mitochondrial DNA. In contrast, in Japanese populations, the etoxazole resistance locus is located on a nuclear chromosome [10]. Hexythiazox resistance is under monogenic control in Australian populations [11], whereas more than one locus is involved in Japanese populations [10]. Similarly, chlorfenapyr resistance in Belgian populations is under polygenic control [12, 13], but Japanese populations might have a resistant allele at only one of the possible chlorfenapyr resistance loci (monogenic) [14]. Moreover, while clofentezine resistance confers a high level of cross-resistance to hexythiazox [15] but not to etoxazole in Australia, hexythiazox resistance probably confers etoxazole resistance in Japanese local populations [10, 16]. Cross-resistance has not been reported between etoxazole and hexythiazox [17, 18].

Monogenic resistance is favored under field selection regimes [19–21], and acaricide resistance is typically monogenic [11, 22–24], although some instances of polygenic resistance have been reported [25–27]. However, each polygenic allele for chlorphenapyr and hexythiazox resistance may be capable of conferring substantial resistance [10, 12, 13]. Previous standard polygenic inheritance descriptions likely cannot adequately predict the evolutionary process of this type of inheritance in which several or a few resistance genes have additive and major effects on the resistant phenotype. In this regard, the LC_{50} for hexythiazox is notably higher for polygenic resistance in Japan (>10,000 mg L^{-1}) [10] than for monogenic resistance in Australia (48 mg L^{-1}) [11], whereas similar LC_{50} values for chlorfenapyr were reported in a Japanese resistant population (2,130 mg L^{-1}) [14] and a Belgian resistant population (2,939 mg L^{-1}) [12]. Moreover, the potential for each resistance gene is individually associated with other distinctive resistance genes.

GENETIC LINKAGE AMONG ACARICIDE RESISTANCE GENES

Most spider mite species have arrhenotokous parthenogenesis; males develop from unfertilized eggs (haploid), and females develop from fertilized eggs (diploid) [28]. This type of parthenogenesis may favor the fixation of fewer but more advantageous alleles [29, 30] but see [31, 32] such as acaricide resistance alleles under acaricide application. Fast fixation of advantageous alleles in male haploids may result in rapid fixation of linked alleles [33], as in the hitchhiking effect [34, 35].

To date, little is known about the genetic linkages among acaricide resistance genes. Uesugi et al. [14] revealed that T. urticae etoxazole and chlorfenapyr resistance genes are genetically linked to each other with a recombination rate of 14.8% (15.3 cM) using the three-point cross with malate-dehydrogenase allozymes [36] and located at 55.8 and 34.3 cM from loci accompanying etoxazole and chlorfenapyr resistance, respectively (Figure 1). They predicted the probability of hitchhiking effects between two closely linked acaricide resistance genes under acaricide selection, calling this phenomenon "apparent cross-resistance." Although hexythiazox resistance is under monogenic control in Australian T. urticae populations, Asahara et al. [10] demonstrated that more than one locus is involved in hexythiazox resistance in a Japanese T. urticae population. Moreover, they found that one of the loci associated with hexythiazox resistance was tightly or completely linked to the etoxazole resistance locus (Figure 1). This suggests that the development of hexythiazox resistance possibly conferred etoxazole resistance in a Japanese local population of T. urticae that had never been exposed to etoxazole application [16].

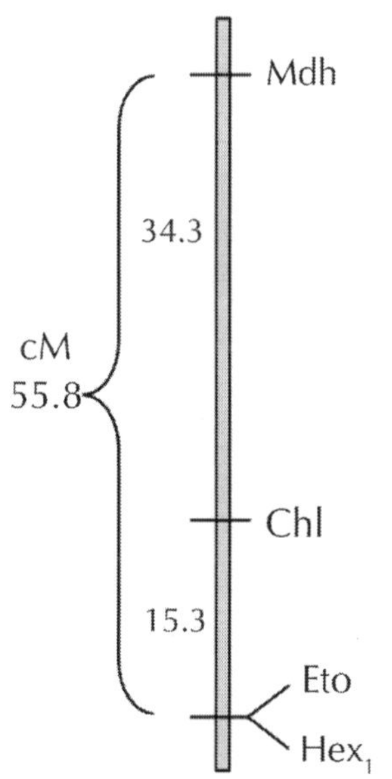

Figure 1: Gene map of a linkage group of loci for malate dehydrogenase (Mdh), resistance to chlorfenapyr (Chl), etoxazole (Eto), and hexythiazox (Hex_1) in T. urticae.

Construction of a genetic linkage map is helpful to determine the effects of genetic relationships among loci associated with distinct acaricides, including the hitchhiking effects during the development of acaricide resistance, as Yan et al. [37] analyzed in yellow fever mosquitoes, Aedes aegypti (L.) (Diptera: Culicidae). Genetic markers, especially microsatellites, should be useful tools for genetic mapping of resistance genes. A substantial number of microsatellites have been cloned and analyzed in spider mites, including the citrus red mite Panonychus citri (McGregor) [38], T. urticae [39–41], and the Kanzawa spider mite Tetranychus kanzawai Kishida [42]. Microsatellites have also been applied to kin species such as Tetranychus turkestani(Ugarov and Nikolskii) [43]. We conduct a linkage analysis of these microsatellites and have revealed two (12 loci) and three (14 loci) linkage groups in T. urticae and T. kanzawai, respectively (Uesugi et al. unpublished data, Matsuo et al. unpublished data). A linkage map with higher resolution should be complete in the near future.

BREEDING PATCH AND GENETIC DIFFERENTIATION

Incidental factors such as gene flow (migration rate) and founder effects affect the dynamics of the local frequency of resistance genes in a metapopulation. Follett and Roderick [44] demonstrated that several resistance foci formed in a metapopulation through model analyses in which migration followed a two-dimensional stepping-stone model under the assumption that individuals move only to neighboring subpopulations in a latticed distribution and a percentage of the populations go extinct at each generation. Spider mites are tiny, wingless arthropods. They exhibit two major dispersal behaviors, crawling and aerial dispersal [45–47]. Crawling is assumed to be used over short distances, such as between leaves within a plant, whereas aerial dispersal may be used for longer distance migration.

We addressed spider mite movement and breeding patches on a host plant because there is some evidence of sibmating in Tetranychus spider mites: a fertilized female deposits many eggs on the same leaf and very little migration occurs during development under uncrowded conditions [48–52]. Hinomoto and Takafuji [53, 54] studied the genetic structure of a T. urticae population on strawberries based on phosphoglucoisomerase allozyme frequency [36]. Interestingly, they found the highest level of genetic differentiation, as measured by the fixation index (F_{ST}) [55], at the smallest habitat (leaflet) level. F_{ST} decreased with increasing the habitat size, that is, leaflet > leaf> plant [53, 54]. In contrast, the value of the inbreeding coefficient (F_{IS}) [55] was the highest at the largest habitat level (subdivided section in the strawberry beds) and the lowest at the smallest habitat level (leaflets) [53]. In this regard, the Wahlund effect probably became larger in larger habitat sizes pooled for F-statistics analyses. Therefore, such genetic structure implies that although T. urticae females and males randomly mate on breeding patches (leaflets), gene flow among the patches is limited. The tendency to form breeding patches probably increases when population densities are low [50, 52] and such genetic structure may often cause genetic differentiation among breeding patches through random genetic drift and/or founder effects.

Adult P. citri females are rather mobile; their microhabitat is greenish twigs with foliage in citrus orchards and the females deposit their eggs on plural leaves [56]. Osakabe and Komazaki [57] focused on the genetic structure in a limited habitat area and performed several laboratory experiments with small citrus seedlings using esterase locus (α-Est1) alleles [58]. They introduced females from two P. citri strains, in which the α-Est1locus was fixed to distinct alleles on the small citrus seedlings, and analyzed phenotype frequencies. As a result, although the inbreeding coefficients within a seedling decreased with population increase, the inbreeding coefficients remained substantially higher than zero even after the mite population had expanded enough to overlap the introduced colonies. Panonychus citri may couple randomly by nature, at least within the smallest breeding patch of a single leaf [57]. Despite their expected mobility [56], adult P. citri females oviposited on very few leaves, at least when the population density was low, and most individuals stayed on the leaves where they had been oviposited during the juvenile developmental stages in the experiments by Osakabe and Komazaki [57].

Adult males usually guard quiescent deutonymphal females and copulate with newly emerged females immediately after their last molts [59]. Once the first copulation for a virgin female has been completed, subsequent mating with other males is ineffective [60, 61]. Many offspring mature on the same leaf, at least during low population density. These spider mite habits might facilitate founder effect in each breeding patch and influence preservation (or fixation) of rare genes, such as acaricide resistance genes.

Under a condition of no acaicide selection, Uesugi et al. [62] demonstrated that genetic drift also works on the dynamics of an acaricide resistance allele; they found a remarkable decrease in mortality following an etoxazole application in two of 32 experimental subpopulations isolated for 13 generations, whereas the resistance did not develop in another 64 subpopulations that experienced limited gene flow.

DISPERSAL DEPENDING ON HABITAT STRUCTURES AND ENVIRONMENTS

Dispersal behavior, colonization, and the resulting population structure of spider mites should strongly affect the distribution and spread of acaricide resistance genes [63]. Because of vigorous reproduction, spider mites cause leaf and plant deterioration, resulting in both food shortage and desiccation [64, 65]. Such unfavorable conditions motivate dispersal behaviors.

Dispersal, which is an important event for settlement of new habitats [63, 66–68], may be affected by the surroundings of microhabitats that may vary depending on the agricultural system, for example, between woody and herbaceous plants and between greenhouse and field situations. This should result in genetic structure diversity of each local population and a consequent geographic distribution of alleles. Goka and Takafuji [69] found that Japanese T. urticae populations on fruit trees and roses could be subdivided into three geographical groups based on allozyme loci allele frequencies, whereas no particular geographical patterns were detectable in populations that inhabited annually herbaceous crops.

Dispersal and Genetic Differentiation in Greenhouses

Most local populations on herbaceous crops occur only temporarily. Thus a population increase may begin by invasion of a small number of mites, facilitating founder effects and genetic differentiation between local populations even if their host plants are adjacently located [69, 70]. Particularly under greenhouse conditions, population growth on herbaceous crops usually starts with a small number of immigrants. They may often experience an intensive bottleneck due to acaricide application in a closed place. This is also the case for greenhouse-grown perennial plants such as roses, although spider mites may maintain their populations on host plants throughout the year. Spider mites are unlikely to use aerial dispersal by means of wind-borne in a greenhouse due to the calm conditions. Windborne dispersal requires adequate wind velocity of 1.5 m/s or more [71–73]. Instead, dispersal by crawling or passive transportation by human movements through agricultural practices is expected in greenhouses [40].

Under such restricted dispersal conditions, genetic differentiation is likely to occur among greenhouse colonies, especially at low population densities [50, 52]. Hinomoto and Takafuji [53] found significant genetic differentiation in a T. urticae population on greenhouse strawberries ($F_{ST\ between\ sites} = 0.16$, $F_{ST\ between\ Plants} = 0.35$). Tsagkarakou et al. [70] investigated population structures in greenhouses in France, where roses, carnations, egg plants, and tomatoes were cultivated. They also found significant differentiation within a greenhouse ($F_{ST\ between\ Sites} =$ 0.01-0.33). In contrast, Navajas et al. [40] obtained low estimates of F_{ST}(0.008-0.09). Nevertheless, the latter findings may not contradict the former examples. Navajas et al. [40] sampled mites from areas incontrovertibly wider than the breeding colony ranges; for example, 15 eggplants were sampled from each site (25 m long transects), and two females per plant were used for the analysis. Such sampling strategies might obscure the effects of genetic differentiation among breeding patches on the population genetic structure data. However, their data demonstrated a general trend toward significant heterozygote deficiency [40], suggesting a Wahlund effect due to substantial isolation of the mite population into smaller breeding patches.

Uesugi et al. [74] analyzed the fine-scaled genetic structure of T. urticae exposed to more than one acaricide heavily applied to rose trees in a greenhouse. They set 18 consecutive quadrats (length: 1.2 m) in two beds (lengths: 22 and 27 m) in a greenhouse and analyzed the genetic structure using microsatellites. As a result, the level of genetic differentiation (θ) [75] among quadrats was higher in the bed with lower mite density (0.221–0.34) than in that with higher mite density (0.134–0.14) [74]. An analysis of molecular variance (AMOVA) revealed a temporal change in allelic frequencies in the bed with lower mite density ($P < .001$), but not in that with higher mite density [74]. These differences likely reflect the larger genetic drift at lower mite density; small population size could cause strong genetic drift, thereby decreasing genetic variation, which is indexed by allelic richness and heterozygosity, and increasing genetic differentiation among quadrat populations [76,77]. Uesugi et al. [74] found that isolation-by-distance [78], calculated by comparing the $F_{ST}/(1-F_{ST})$ values with the spatial distance of each quadrat population [79], was significant in the bed with higher mite density (Mantel test, $P < .05$) but not in that with lower mite density. A significant positive autocorrelation was detected by spatial autocorrelation analysis [80] only within a short range (2.4–3.6 m) of the bed with higher mite density, suggesting that migration was limited to a short range, whereas larger-scale dispersal, such as aerial dispersal, did not appear to contribute to the genetic structure [74].

Spider mites within a breeding patch are likely to mate randomly rather than assortatively, and an increase in population density effectively reduces F_{ST} values among patches [54, 57]. Nevertheless, fine-scale genetic structure is preserved under stable environments such as a perennial host plant in a greenhouse [74]. Yano [81] demonstrated collective dispersal and resulting aggregation at a new feeding site in T. urticae; crawling dispersers tended to follow the trails of former dispersers, which may be associated with the threads spun by spider mites. Therefore, the tendency for high F_{IS} values in greenhouse spider mites may be attributed to the traits of spider mites forming breeding patches on small habitats, such as an individual leaf, and the limitation of immigration among patches. Consequently, gene flow between breeding patches is delayed even if mites are located on the same plant. Instead, founder effects and genetic drift promote genetic differentiation between breeding patches via fixation to specific alleles in each breeding patch.

Aerial Dispersal Mode and Gene Flow in Orchards and Citrus Groves

Spider mites aerially disperse by means of wind-borne or ballooning. Young gravid adult females are the dominant dispersers in both aerial dispersal means [82, 83]. However, species that exploit the wind-borne and ballooning probably achieve different arrival distances. In a practical investigation, dispersal of only 10–20 m occurred from blackberry into corn [66, 71]. Model analyses have suggested 5.5–12 m and 16–48 m dispersal from corn and apple, respectively [84]. In peanut field experiments by Boykin and Campbell [71], regardless of release height for aerial dispersal, most T. urticae caught by sticky traps were only 60 cm above the ground (the lowest traps) and catch numbers decreased as the traps were set higher (the highest traps were set 2.1 m above the ground). Therefore, as a general rule for the passive aerial ground dispersal, the number of arrivals per unit area probably declines steeply with increasing distance from the point of departure [85]. In contrast, in the field experiments by Fleschner et al. [82], ballooning O. punicae were captured at the top 60 cm of a 6 m (5 ×5 cm) pole in an avocado grove, suggesting that they drifted upward after takeoff and that mites dispersing by ballooning migrate longer distances than do windborne mites. In fact, P. citri has been observed aerially migrating between adjacent citrus groves [82] and from distant citrus groves to Japanese pear orchards [86]. Almost imperceptible air currents are sufficient for mites to balloon [82].

Uesugi et al. [87] investigated the structures of T. urticae populations in two apple orchards using five microsatellite loci. In the first orchard, apple trees were planted in lines at intervals of 5–8 m, and their branches did not contact neighboring trees. In the second orchard, apple trees were planted at intervals of 2 and 7.5 m within and between lines, respectively. To exclude effects on gene flow from direct crawling dispersal between apple trees, Uesugi et al. sampled mites from apple trees along a diagonal transect in the second orchard. Ultimately, mites were collected from apple trees separated by distances of 10–24 m within about 100 m of the lines in these orchards. As a result, heterozygote deficiency (f) [75] and θ among trees were positive in these orchards ($f = 0.328$ and $\theta = 0.028$ in the first orchard, and $f = 0.078$ and $\theta = 0.012$ in the second orchard). This suggests that there was a certain level of limitation to mite migration, and complete genetic homogenization was prevented within a scale of the studied sites. Nevertheless, analyses of genetic similarity and spatial location using dendrograms, isolation-by-distance, and multiallelic correlograms revealed no significant genetic differentiation among trees in these orchards. This may be explained that the level of gene flow is relatively high over long distances of $< 100\,m$, resulting in no genetic structure among mite populations in distant trees. Similarly, Grafton-Cardwell et al. [63] found that the frequencies of T. urticae and Tetranychus pacificus McGregor resistant to cyhexatin and dicofol were equivalent in on neighboring almonds and cotton, although cyhexatin had been applied only on almonds while dicofol had been applied only on cotton, indicating the occurrence of substantial gene flow between mite populations on these plants by aerial dispersal. Notwithstanding, significant isolation-by-distance was found in an analysis of wider geographic scales (populations were 10 m to more than 100 km apart) using allozymes in Greece [88]. Those suggest limitations for this mite to regionally disperse using the wind.

Genetic differentiation among geographically distant populations was also found in P. citri. Osakabe et al. [89] analyzed P. citri genetic structure hierarchically from the level between patches within a tree to between regions (Honshu and Kyushu islands of Japan), using the coefficient of gene differentiation (G_{ST}) [90, 91]. They found that G_{ST} tended to be smaller with decreasing area and distance between populations [89]. This geographic trend was consistent with the results of a protein analysis by Osakabe and Sakagami [92]. However, the diversity of P. citri populations within a local population and/or a microhabitat was different from that of T. urticae, which maintained frequent gene flow in a local population [87, 93]. The divergence within a grove and/or a tree clearly accounted for the total diversity of P. citri populations; G_{ST} values between trees within a grove and between patches within a tree were larger than those between groves. These results suggest that P. citri frequently migrates between citrus groves within a locality but not across wider areas. Moreover, higher diversity among trees (or patches) within groves than between groves implies that this mite migrates mainly by aerial drift (ballooning) between groves [82], whereas crawling dispersal is limited. Thus interbreeding between patches is infrequent and solely dependent on interpatch distance [57].

Consequently, the two modes of aerial dispersal in spider mites, wind-borne and ballooning, likely result in widely different population structures. Dispersion patterns of species exploiting the wind-borne might be explained using an aerodynamic model, such as the seed flux model [94, 95], as described by Jung and Croft [84]. These patterns may represent a practical decline in the number of arrivals per unit area with increasing distance from departure point [85]. The significance of microhabitat for P. citri genetic population structures was also demonstrated in marine organisms with reduced dispersal, such as polychaetes, which disperse only during the planktonic larval stage [96].

EFFECTS OF DISPERSAL AND HABITAT STABILITY ON ACARICIDE RESISTANCE DEVELOPMENT

Patch size and persistence of host plants may largely influence population biology as well as gene flow between patches (colonies) and/or local populations. Additionally, variation in the initial frequency of pesticide resistance genes and intensity of selection force by pesticide application may result in different pesticide resistance development rates among local pest populations [97], which may need to be considered to determine the velocity of acaricide resistance development and prevalence [19, 98–100].

As described above, T. urticae populations inhabiting stable woody plant habitats (fruit trees and roses) were geographically differentiated into genetically structured groups in Japan, but such differentiation did not occur in populations inhabiting annual herbaceous crops [69]. Goka [101] assessed resistance to two acaricides, pyridaben and fenpyroximate (used in Japan since 1991), in T. urticae populations collected from 48 deciduous fruit tree orchards and rose gardens and from 42 patches of herbaceous plants (some of which had been grown in greenhouses) from spring 1993 to autumn 1994. They found a rapid development of acaricide resistance in the woody plant populations. Only 7 and 8 of 48 populations showed 100% mortality after the application of pyridaben (200 mg l^{-1}) and fenpyroximate (50 mg l^{-1}), respectively. In contrast, resistant individuals were found in more than 80% of the populations; more than half of the tested individuals were alive in 24 and 10 populations after the application of pyridaben and fenpyroximate, respectively. Remarkably, no individual died by either acaricide in six populations (Figures 2(a) and 2(b)). In contrast, resistant individuals were found in a few populations after application of the acaricides, and more than 50% of individuals survived in an identical population (Figures 2(c) and 2(d); arrows). Different velocities of resistance development between these habitats were not caused by differences in the frequency of acaricide application (Figure 3). Instead, these phenomena may have occurred due to differences in mite population size which can be harbored in the habitat and habitat stability. Mite populations on woody plants are stable residents such that gene flow between them results in the strong likelihood of selection and accumulation of rare genes for resistance, whereas those on herbaceous crops are unstable transients such that rare genes would be unlikely to be retained for long periods [101]. Yet, resistance to the acaricides was obviously developed in a single population on herbaceous plants (Figure 3; arrows), which could have resulted from founder effects.

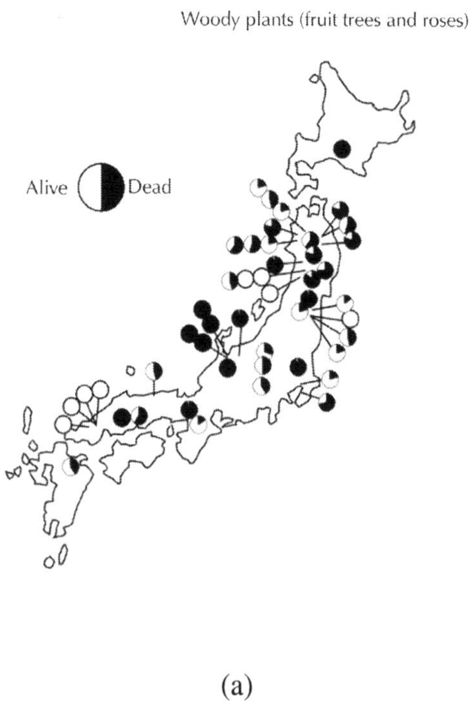

(a)

(b)

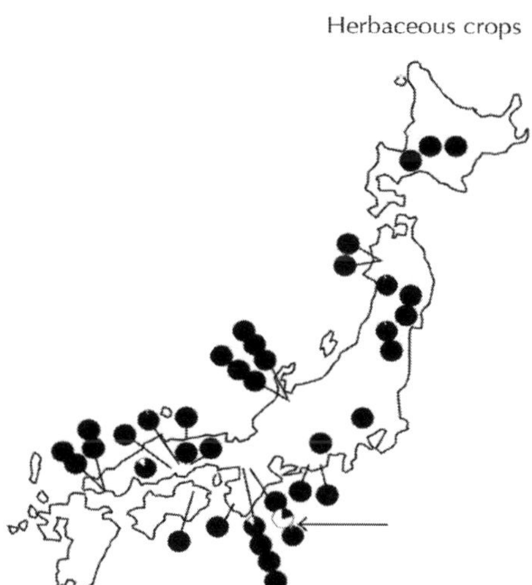

(c)

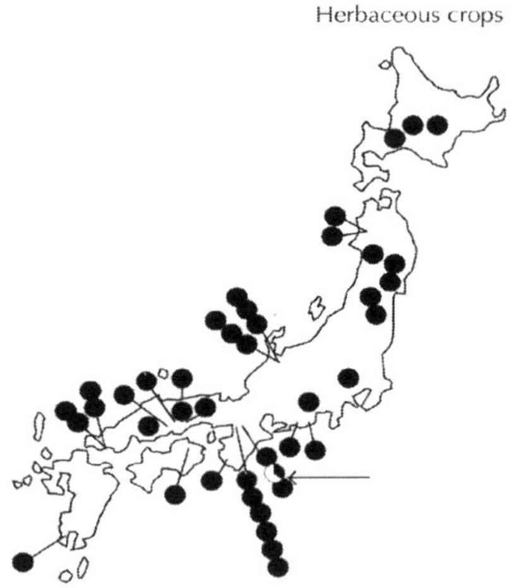

(d)

Figure 2: Geographic variation of susceptibility to acaricides in 48 and 42 T. urticaepopulations collected from woody plants (fruit trees and rose plants) (a), (b) and herbaceous crops (c), (d).

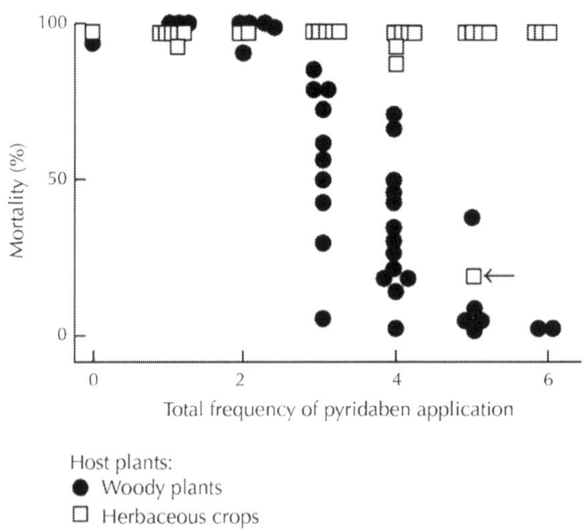

(a)

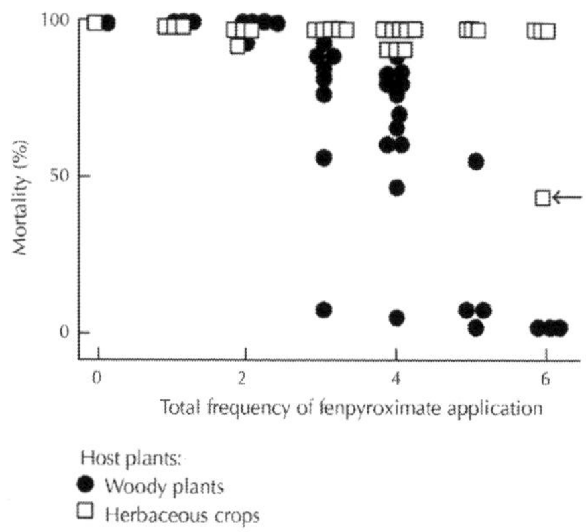

(b)

Figure 3: Correlation between total field applications of acaricides (200 mg L^{-1} pyridaben; 50 mg L^{-1} fenpyroximate) and mortality for local populations.

In the assessment by Osakabe et al. [89], P. citri showed higher genetic diversity in acaricide resistance genes within a grove and within a tree than between groves; a two-level nested analysis of variance revealed significant difference in fenpyroximate resistance among trees within groves ($P < .025$) but not among groves ($P > .25$). This may reflect the population structure revealed using gene markers, as described above. Similar trends also occurred in the distribution of etoxazole resistance [89]. A variation in fenpyroximate susceptibility was found even among patches within a tree. In a field study by Yamamoto et al. [102], strong spatial aggregation of the hexythiazox resistance gene in P. citri was identified both between trees and within a single tree in a citrus grove. This result suggests that acaricide resistance genes are spread based on the population structure, that is, limited crawling dispersal within foliage [57] resulting in between-patch diversity and frequent longer aerial dispersal by ballooning resulting in between-grove homogeneity.

A study of genetic differentiation of a T. urticae population on roses in a greenhouse estimated short dispersal distances and thus low spread velocity of acaricide resistance genes [74]. Nevertheless, milbemectin resistance genes, which had been distributed only partially in the population in July 2006, spread over the bed after only

2 months (Uesugi et al. unpublished data). During those 2 months, the roses were sprayed twice with milbemectin. The mechanism of such quick resistance development is unclear. Given that gene flow among patches within a bed did not occur frequently, an increase in the gene frequencies as effects of selection likely functioned to develop resistance. As Navajas et al. [40] noted, human-promoted passive transportation through agricultural practices might facilitate the quick spread of resistance genes.

COMPREHENSIVE ASPECTS OF ACARICIDE RESISTANCE EVOLUTION

The initial frequency of resistance genes and the gene dominance are influential factors that determine the rate of resistance development [44]. In general, given a nonstructured population, mutated alleles associated with pesticide resistance likely exist at very low frequencies, and so in most of the resistance alleles are expected to be heterozygotes. Nevertheless, theoretical model experiments have revealed that in a metapopulation condition even lower-frequency alleles form concentrated foci in a set of subpopulations with certain qualifications for selection of pesticides, extinction of subpopulations, and migration [44]. With these qualifications, subpopulation extinction and subsequent migration and reconstruction possibly occur based on incidental changes in gene frequency through founder effects. This sometimes results in the construction of subpopulations in which the resistance allele frequencies are relatively higher, making selection by pesticides more effective for resistance development.

Data concerning genetic structure suggest that spider mite populations are metapopulations. Although Uesugi et al. [87] did not find differentiation within an apple orchard, there was significant heterozygote deficiency (significantly positive F_{IS} value) in all trees, suggesting that Wahlund effects may have been caused by the division of the population into small breeding patches. Young gravid females are the main dispersers of spider mites when they experience high density [83]. Whether they disperse aerially or crawl, the dispersers may arrive alone on a new feeding site. These dispersal behaviors and restricted dispersal ability by crawling imply that founder effects frequently act on spider mite colonies, promoting the development of resistance foci through selection for acaricides. The resistance genes will then spread limitedly to neighboring colonies by gene flow or colony expansion by propagation. Moreover, the resistance genes may be spread from foci to other distant habitats by aerial dispersal in each species.

We have no data about the probability of new-linkage formation among resistance genes controlled by distinctive loci on the same chromosome by recombination in field populations. The frequent occurrence of resistance foci may increase the mating chances between mites bearing different resistance genes and might be followed by the occurrence of hitchhiking effects. Precise genetic experiments will be necessary to elucidate this issue.

ACARICIDE RESISTANCE MANAGEMENT AND GLOBALIZATION

Species identification is a basic issue for biosecurity as injurious species must be identified to prevent their invasion and establishment in new areas [103–105]. Eight spider mite species including T. urticae, T. kanzawai, P. citri, and P. ulmi are serious cosmopolitan pests and are now listed as nonquarantine pests in the Plant Protection Law of Japan. Therefore, the risk of introducing exotic resistance alleles to inland areas is undoubtedly increasing. Even if there are inland populations resistant to the same acaricide, an invasion of an exotic resistance gene may be a potential risk over the long term because an introduced resistance gene would not be identical to that of inland populations. Therefore, more studies on the diversity of acaricide resistance genes should be performed using genetic marker linkage maps. Moreover, methods to control the spread of such exotic resistance genes may be important to prevent the development of more complex resistance, and mechanisms of regional dispersal and resulting gene flow may have to be addressed.

PERSPECTIVES

Identifying a mechanism for the development of pesticide resistance is important for advancing pesticide resistance management for arthropod pests. In spider mites, past genetic and ecological studies have comprehensively suggested that the local concentration of resistance genes (increasing gene frequency in breeding patches) resulting from genetic diversity within habitats based on their biological traits and selection by acaricides and gene flow from selection sites to surroundings (local and/or regional spread of resistance) are the processes of acaricide resistance evolution.

Our study has revealed the consequences of habitat stability to acaricide resistance development [101] and implies that disturbing the expansion of incipient colonies is effective for restraining acaricide resistance development and avoiding complications with spider mites. Although natural enemies such as phytoseiid mites (Acari: Phytoseiidae) play the role of the disturber, their settlement and predation activity are important. We expect the development of new technologies for the management of natural enemies, such as artificial diet provisions [106]. Other factors possibly disturbing incipient spider mite colonies such as ultraviolet rays, which are lethal to spider mites [107], should also be exploited. Overall, the establishment of an integrated mite management strategy aimed at lowering spider mite population density may be the best way to manage acaricide resistance.

REFERENCES

1. M. E. Whalon, D. Mota-Sanchez, and R. M. Hollingworth, "Analysis of global pesticide resistance in arthropods," in Global Pesticide Resistance in Arthropods, M. E. Whalon, D. Mota-Sanchez, and R. M. Hollingworth, Eds., pp. 5–31, CABI, Oxfordshire, UK, 2008.
2. R. M. Hollingworth and K. Dong, "The biochemical and molecular genetic basis of resistance to pesticides in arthropods," in Global Pesticide Resistance in Arthropods, M. E. Whalon, D. Mota-Sanchez, and R. M. Hollingworth, Eds., pp. 40–89, CABI, Oxfordshire, UK, 2008.
3. Y.-S. Lee, M.-H. Song, K.-S. Ahn, K.-Y. Lee, J.-W. Kim, and G.-H. Kim, "Monitoring of acaricide resistance in two-spotted spider mite (Tetranychus urticae) populations from rose greenhouses

in Korea," Journal of Asia-Pacific Entomology, vol. 6, no. 1, pp. 91–96, 2003.

4. T. Van Leeuwen, S. Van Pottelberge, and L. Tirry, "Comparative acaricide susceptibility and detoxifying enzyme activities in field-collected resistant and susceptible strains of Tetranychus urticae," Pest Management Science, vol. 61, no. 5, pp. 499–507, 2005.
5. D. J. Pree, K. J. Whitty, and L. Van Driel, "Baseline susceptibility and cross resistances of some new acaricides in the European red mite, Panonychus ulmi," Experimental and Applied Acarology, vol. 37, no. 3-4, pp. 165–171, 2005.
6. G. P. Georghiou and C. E. Taylor, "Genetic and biological influences in the evolution of insecticide resistance," Journal of Economic Entomology, vol. 70, no. 3, pp. 319–323, 1977.
7. G. P. Georghiou and C. E. Taylor, "Operational influences in the evolution of insecticide resistance,"Journal of Economic Entomology, vol. 70, no. 5, pp. 653–658, 1977.
8. I. Denholm and M. W. Rowland, "Tactics for managing pesticide resistance in arthropods: theory and practice," Annual Review of Entomology, vol. 37, no. 1, pp. 91–112, 1992.
9. S. Lee, K. Ahn, C. Kim, S. Shin, and G. Kim, "Inheritance and stability of etoxazole resistance in twospotted spider mite, Tetranychus urticae, and its cross resistance," Korean Journal of Applied Entomology, vol. 43, pp. 43–48, 2004.
10. M. Asahara, R. Uesugi, and Mh. Osakabe, "Linkage between one of the polygenic hexythiazox resistance genes and an etoxazole resistance gene in the twospotted spider mite (Acari: Tetranychidae)," Journal of Economic Entomology, vol. 101, no. 5, pp. 1704–1710, 2008.
11. G.A. Herron and J. Rophail, "Genetics of hexythiazox resistance in two spotted spider mite, Tetranychus urticae Koch," Experimental and Applied Acarology, vol. 17, no. 6, pp. 423–431, 1993.
12. T. Van Leeuwen, V. Stillatus, and L. Tirry, "Genetic analysis and cross-resistance spectrum of a laboratory-selected chlorfenapyr resistant strain of two-spotted spider mite (Acari: Tetranychidae),"Experimental and Applied Acarology, vol. 32, no. 4, pp. 249–261, 2004.
13. T. Van Leeuwen, S. Van Pottelberge, and L. Tirry, "Biochemical analysis of a chlorfenapyr-selected resistant strain of Tetranychus

urticae Koch," Pest Management Science, vol. 62, no. 5, pp. 425–433, 2006.

14. R. Uesugi, K. Goka, and Mh. Osakabe, "Genetic basis of resistances to chlorfenapyr and etoxazole in the two-spotted spider mite (Acari: Tetranychidae)," Journal of Economic Entomology, vol. 95, no. 6, pp. 1267–1274, 2002.
15. G. Herron, V. Edge, and J. Rophail, "Clofentezine and hexythiazox resistance in Tetranychus urticaeKoch in Australia," Experimental and Applied Acarology, vol. 17, no. 6, pp. 433–440, 1993.
16. M. Kobayashi, S. Kobayashi, and T. Nishimori, "Occurrence of etoxazole resistance individuals of the two-spotted spider mite, Tetranychus urticae Koch from a limited region," Japanese Journal of Applied Entomology and Zoology, vol. 45, no. 2, pp. 83–88, 2001 (Japanese
17. T. Ishida, J. Suzuki, and Y. Tsukidate, "YI-5301, a novel oxazoline acaricide," in Proceedings of the British Crop Protection Conference on Pests & Diseases, vol. 2, pp. 37–44, 1994.
18. R. Nauen and G. Smagghe, "Mode of action of etoxazole," Pest Management Science, vol. 62, no. 5, pp. 379–382, 2006.
19. R. T. Roush and J. A. McKenzie, "Ecological genetics of insecticide and acaricide resistance," Annual Review of Entomology, vol. 32, pp. 361–380, 1987.
20. J. A. McKenzie, A. G. Parker, and J. L. Yen, "Polygenic and single gene responses to selection for resistance to diazinon in Lucilia cuprina," Genetics, vol. 130, no. 3, pp. 613–620, 1992.
21. Y. Carrière and D. A. Roff, "Change in genetic architecture resulting from the evolution of insecticide resistance: a theoretical and empirical analysis," Heredity, vol. 75, no. 6, pp. 618–629, 1995.
22. T. E. Martinson, T. J. Dennehey, J. P. Nyrop, and W. H. Reissig, "Field-measurements of selection for twospotted spider mite (Acari: Tetranychidae) resistance to dicofol in apple orchards," Journal of Economic Entomology, vol. 84, pp. 7–16, 1991.
23. A. Yamamoto, H. Yoneda, R. Hatano, and M. Asada, "Genetic analysis of hexythiazox resistance in the citrus red mite, Panonychus citri (McGregor)," Journal of Pesticide Science, vol. 20, pp. 513–519, 1995.

24. K. Goka, "Mode of inheritance of resistance to three new acaricides in the Kanzawa spider mite, Tetranychus kanzawai Kishida (Acari: Tetranychidae)," Experimental and Applied Acarology, vol. 22, no. 12, pp. 699–708, 1998
25. A. Mizutani, F. Kumayama, K. Ohba, T. Ishiguro, and Y. Hayashi, "Inheritance of resistance to cyhexatin in the Kanzawa spider mite, Tetranychus kanzawai Kishida (Acarina: Tetranychidae)," Applied Entomology and Zoology, vol. 23, pp. 251–255, 1988.
26. M. A. Keena and J. Granett, "Genetic analysis of propargite resistance in Pacific spider mites and twospotted spider mites (Acari: Tetranychidae)," Journal of Economic Entomology, vol. 83, pp. 655–661, 1990.
27. J. M. Clark, J. G. Scott, F. Campos, and J. R. Bloomquist, "Resistance to avermectins: extent, mechanisms, and management implications," Annual Review of Entomology, vol. 40, pp. 1–30, 1995.
28. W. Helle and L. P. Pijnacker, "Parthenogenesis, chromosomes and sex," in Spider Mites. Their Biology, Natural Enemies and Control. Vol. 1A, W. Helle and M. W. Sabelis, Eds., pp. 129–139, Elsevier, Amsterdam, The Netherlands, 1985.
29. L. J. Lester and R. K. Selander, "Population genetics of haplodiploid insects," Genetics, vol. 92, no. 4, pp. 1329–1345, 1979.
30. R. H. Crozier, "Adaptive consequences of male-haploidy," in Spider Mites. Their Biology, Natural Enemies and Control. Vol. 1A, W. Helle and M. W. Sabelis, Eds., pp. 201–222, Elsevier, Amsterdam, The Netherlands, 1985.
31. Y. Carrière, "Haplodiploidy, sex, and the evolution of pesticide resistance," Journal of Economic Entomology, vol. 96, no. 6, pp. 1626–1640, 2003.
32. D. W. Crowder, Y. Carrière, B. E. Tabashnik, P. C. Ellsworth, and T. J. Dennehy, "Modeling evolution of resistance to pyriproxyfen by the sweetpotato whitefly (Homoptera: Aleyrodidae)," Journal of Economic Entomology, vol. 99, no. 4, pp. 1396–1406, 2006.
33. P. Pamilo and R. H. Crozier, "Genetic variation in male haploids under deterministic selection," Genetics, vol. 98, pp. 199–214, 1981.

34. K. Kojima and H. E. Schaffer, "Survival process of linked mutant genes," Evolution, vol. 21, pp. 518–538, 1967.
35. J. Maynard-Smith and J. Haigh, "The hitch-hiking effect of a favorable gene," Genetics Research, vol. 23, pp. 23–35, 1974.
36. K. Goka and A. Takafuji, "Genetic basis of esterase, malate dehydrogenase and phosphoglucoisomerase allozymes in the two-spotted spider mite, Tetranychus urticae Koch," Applied Entomology and Zoology, vol. 30, pp. 529–535, 1995.
37. G. Yan, D. D. Chadee, and D. W. Severson, "Evidence for genetic hitchhiking effect associated with insecticide resistance in Aedes aegypti," Genetics, vol. 148, no. 2, pp. 793–800, 1998.
38. Mh. Osakabe, N. Hinomoto, S. Toda, S. Komazaki, and K. Goka, "Molecular cloning and characterization of a microsatellite locus found in an RAPD marker of a spider mite, Panonychus citri(Acari: Tetranychidae)," Experimental and Applied Acarology, vol. 24, no. 5-6, pp. 385–395, 2000.
39. M. J. Navajas, H. M. A. Thistlewood, J. Lagnel, and C. Hughes, "Microsatellite sequences are under-represented in two mite genomes," Insect Molecular Biology, vol. 7, no. 3, pp. 249–256, 1998.
40. M. Navajas, M. J. Perrot-Minnot, J. Lagnel, A. Migeon, T. Bourse, and J. M. Cornuet, "Genetic structure of a greenhouse population of the spider mite Tetranychus urticae: spatio-temporal analysis with microsatellite markers," Insect Molecular Biology, vol. 11, no. 2, pp. 157–165, 2002.
41. R. Uesugi and Mh. Osakabe, "Isolation and characterization of microsatellite loci in the two-spotted spider mite, Tetranychus urticae (Acari: Tetranychidae)," Molecular Ecology Notes, vol. 7, no. 2, pp. 290–292, 2007.
42. S. Nishimura, N. Hinomoto, and A. Takafuji, "Isolation, characterization, inheritance and linkage of microsatellite markers in Tetranychus kanzawai (Acari: Tetranychidae)," Experimental and Applied Acarology, vol. 31, no. 1-2, pp. 93–103, 2003.
43. X. Bailly, A. Migeon, and M. Navajas, "Analysis of microsatellite variation in the spider mite pestTetranychus turkestani (Acari: Tetranychidae) reveals population genetic structure and raises questions about related ecological factors," Biological Journal of the Linnean Society, vol. 82, no. 1, pp. 69–78, 2004.

44. P. A. Follett and G. K. Roderick, "Adaptation to insecticides in Colorado potato beetle: single- and meta-population models," in Chrysomelidae Biology, P. H. A. Jolivet and M. L. Cox, Eds., vol. 4, pp. 289–302, SPB Academic, Amsterdam, The Netherlands, 1996.
45. L. Jeppson, H. H. Keifer, and E. W. Baker, Mites Injurious to Economic Plants, University of California Press, Berkeley, Calif, USA, 1975.
46. G. G. Kennedy and D. R. Smitley, "Dispersal," in Spider Mites. Their Biology, Natural Enemies and Control. Vol. 1A, W. Helle and M. W. Sabelis, Eds., pp. 233–242, Elsevier, Amsterdam, The Netherlands, 1985.
47. Mh. Osakabe, H. Isobe, A. Kasai, R. Masuda, S. Kubota, and M. Umeda, "Aerodynamic advantages of upside down take-off for aerial dispersal in Tetranychus spider mites," Experimental and Applied Acarology, vol. 44, no. 3, pp. 165–183, 2008.
48. W. Helle, "Genetics of resistance to organophophorus compounds and its relation to diapause inTetranychus urticae Koch," Tijdschrift over Plantenziekten, vol. 68, pp. 1–41, 1962.
49. N. W. Hussey and W. J. Parr, "Dispersal of the glasshouse red spider mite Tetranychus urticae Koch (Acarina, Tetranychidae)," Entomologia Experimentalis et Applicata, vol. 6, no. 3, pp. 207–214, 1963.
50. G. Nachman, "Temporal and spatial dynamics of an acarina predator-prey system," Journal of Animal Ecology, vol. 50, pp. 433–451, 1981.
51. A. Kondo and A. Takafuji, "Resource utilization pattern of two species of tetranychid mites (Acarina: Tetranychidae)," Researches on Population Ecology, vol. 27, no. 1, pp. 145–157, 1985.
52. P.-M. So, "Distribution patterns of and sampling plans for Tetranychus urticae Koch (Acarina: Tetranychidae) on roses," Researches on Population Ecology, vol. 33, no. 2, pp. 229–243, 1991.
53. N. Hinomoto and A. Takafuji, "Studies on the population structure of the two-spotted spider mite,Tetranychus urticae Koch, by allozyme variability analysis," Applied Entomology and Zoology, vol. 29, no. 2, pp. 259–266, 1994.

54. N. Hinomoto and A. Takafuji, "Genetic changes in the population structure of the two-spotted spider mite, Tetranychus urticae Koch (Acari: Tetranychidae) on vinyl-house strawberries," Applied Entomology and Zoology, vol. 30, pp. 521–528, 1995.
55. S. Wright, Evolution and the Genetics of Populations. Vol. 4. Variability within and among Natural Popuolations, The University of Chicago Press, Chicago, Ill, USA, 1978.
56. K. Wanibuchi andY. Saito, "The process of population increase and patterns of resource utilization of two spider mites, Oligonychus ununguis (Jacobi) and Panonychus citri (McGregor), under experimental conditions (Acari: Tetranychidae)," Researches on Population Ecology, vol. 25, no. 1, pp. 116–129, 1983.
57. Mh. Osakabe and S. Komazaki, "Laboratory experiments on a change in genetic structure with an increase of population density in the citrus red mite population, Panonychus citri (McGregor) (Acari: Tetranychidae)," Applied Entomology and Zoology, vol. 34, no. 4, pp. 413–420, 1999.
58. Mh. Osakabe, "Genetic analysis of esterases in the citrus red mite, Panonychus citri (McGregor) (Acari: Tetranychidae)," Applied Entomology and Zoology, vol. 26, pp. 307–312, 1991.
59. W. Helle and W. P. J. Overmeer, "Variability in tetranychid mites," Annual Review of Entomology, vol. 18, pp. 97–120, 1973.
60. D. A. Potter and D. L. Wrensch, "Interrupted matings and the effectiveness of second inseminations in the twospotted spider mite," Annals of the Entomological Society of America, vol. 71, pp. 882–885, 1978.
61. A. Takafuji, "Effectiveness of second mating for two incompatible types of the citrus red mite,Panonychus citri (McGregor)," Researches on Population Ecology, vol. 28, no. 1, pp. 91–101, 1986.
62. R. Uesugi, K. Goka, and Mh. Osakabe, "Development of genetic differentiation and postzygotic isolation in experimental metapopulations of spider mites," Experimental and Applied Acarology, vol. 31, no. 3-4, pp. 161–176, 2003.
63. E. E. Grafton-Cardwell, J. Granett, and S. M. Normington, "Influence of dispersal from almonds on the population dynamics and acaricide resistance frequencies of spider mites infesting

neighboring cotton,"Experimental and Applied Acarology, vol. 10, no. 3-4, pp. 187–212, 1991.

64. F. V. Sances, J. A. Wyman, and I. P. Ting, "Physiological responses to spider mite infestations on strawberries," Environmental Entomology, vol. 8, pp. 711–714, 1979.
65. J. D. De Angelis, K. C. Larson, R. E. Berry, and G. W. Krantz, "Effects of spider mite injury on transpiration and leaf water status in peppermint," Environmental Entomology, vol. 11, pp. 975–978, 1982.
66. R. L. Brandenburg and G. G. Kennedy, "Intercrop relationships and spider mite dispersal in a corn/peanut agro-ecosystem," Entomologia Experimentalis et Applicata, vol. 32, no. 3, pp. 269–276, 1982.
67. D. R. Smitley and G. G. Kennedy, "Aerial dispersal of the two-spotted spider mite (Tetranychus urticae) from field corn," Experimental and Applied Acarology, vol. 5, no. 1-2, pp. 33–46, 1988.
68. D. S. Lawson, J. P. Nyrop, and T. J. Dennehy, "Aerial dispersal of European red mites (Acari: Tetranychidae) in commercial apple orchards," Experimental and Applied Acarology, vol. 20, no. 4, pp. 193–202, 1996.
69. K. Goka and A. Takafuji, "Allozyme variations among populations of the two-spotted spider mite,Tetranychus urticae Koch, in Japan," Applied Entomology and Zoology, vol. 30, pp. 567–579, 1995.
70. A. Tsagkarakou, M. Navajas, F. Rousset, and N. Pasteur, "Genetic differentiation in Tetranychus urticae(Acari: Tetranychidae) from greenhouses in France," Experimental and Applied Acarology, vol. 23, no. 5, pp. 365–378, 1999.
71. L. S. Boykin and W. V. Campbell, "Wind dispersal of the twospotted spider mite (Acari: Tetranychidae) in North Carolina peanut fields," Environmental Entomology, vol. 13, pp. 221–227, 1984.
72. D. R. Smitley and G. G. Kennedy, "Photo-oriented aerial-dispersal behavior of Tetranychus urticae(Acari: Tetranychidae) enhances escape from the leaf surface," Annals of the Entomological Society of America, vol. 78, pp. 609–614, 1985.

73. D. C. Margolies, “Conditions eliciting aerial dispersal behavior in banks grass mite, Oligonychus pratensis (Acari: Tetranychidae),” Environmental Entomology, vol. 16, pp. 928–932, 1987.
74. R. Uesugi, Y. Kunimoto, and Mh. Osakabe, “The fine-scale genetic structure of the two-spotted spider mite in a commercial greenhouse,” Experimental and Applied Acarology, vol. 47, no. 2, pp. 99–109, 2009.
75. B. S. Weir and C. C. Cockerham, “Estimating F-statistics for the analysis of population structure,”Evolution, vol. 38, no. 6, pp. 1358–1370, 1984.
76. P. A. Fuerst and T. Maruyama, “Considerations on the conservation of alleles and of genic heterozygosity in small managed populations,” Zoo Biology, vol. 5, pp. 171–180, 1986.
77. M. Gilpin, “The genetic effective size of a metapopulation,” Biological Journal of the Linnean Society, vol. 42, no. 1-2, pp. 165–175, 1991.
78. S. Wright, “Isolation by distance,” Genetics, vol. 28, pp. 114–138, 1943.
79. F. Rousset, “Genetic differentiation and estimation of gene flow from F-statistics under isolation by distance,” Genetics, vol. 145, no. 4, pp. 1219–1228, 1997.
80. P. E. Smouse and R. Peakall, “Spatial autocorrelation analysis of individual multiallele and multilocus genetic structure,” Heredity, vol. 82, no. 5, pp. 561–573, 1999.
81. S. Yano, “Collective and solitary behaviors of twospotted spider mite (Acari: Tetranychidae) are induced by trail following,” Annals of the Entomological Society of America, vol. 101, no. 1, pp. 247–252, 2008.
82. C. A. Fleschner, M. E. Badgley, D. W. Ricker, and J. C. Hall, “Air drift of spider mites,” Journal of Economic Entomology, vol. 49, pp. 624–627, 1956.
83. R. Mitchell, “Growth and population dynamics of a spider mite (Tetranychus urticae K., Acarina: Tetranychidae),” Ecology, vol. 54, pp. 1349–1355, 1973.
84. C. Jung and B. A. Croft, “Aerial dispersal of phytoseiid mites (Acari: Phytoseiidae): estimating falling speed and dispersal distance of adult females,” Oikos, vol. 94, no. 1, pp. 182–190, 2001.

85. D. H. Janzen, "Herbivores and the number of tree species in tropical forests," The American Naturalist, vol. 104, pp. 501–528, 1970.
86. A. Takafuji and N. Morimoto, "Diapause attributes and seasonal occurrences of two populations of the citrus red mite, Panonychus citri (McGregor) on pear (Acarina: Tetranychidae)," Applied Entomology and Zoology, vol. 18, pp. 525–532, 1983.
87. R. Uesugi, T. Sasawaki, and Mh. Osakabe, "Evidence of a high level of gene flow among apple trees inTetranychus urticae," Experimental and Applied Acarology, vol. 49, no. 4, pp. 281–290, 2009.
88. A. Tsagkarakou, M. Navajas, P. Papaioannou-Souliotis, and N. Pasteur, "Gene flow among Tetranychus urticae (Acari: Tetranychidae) populations in Greece," Molecular Ecology, vol. 7, no. 1, pp. 71–79, 1998.
89. Mh. Osakabe, K. Goka, S. Toda, T. Shintaku, and H. Amano, "Significance of habitat type for the genetic population structure of Panonychus citri (Acari: Tetranychidae)," Experimental and Applied Acarology, vol. 36, no. 1, pp. 25–40, 2005.
90. M. Nei, "Analysis of gene diversity in subdivided populations," Proceedings of the National Academy of Sciences of the United States of America, vol. 70, no. 12, pp. 3321–3323, 1973.
91. R. Chakraborty, "Gene-diversity analysis in nested subdivided populations," Genetics, vol. 96, pp. 721–726, 1980.
92. Mh. Osakabe and Y. Sakagami, "Protein differences detected by two-dimensional electrophoresis among local populations of Panonychus citri (McGregor) (Acari: Tetranychidae) in Japan," Applied Entomology and Zoology, vol. 28, no. 4, pp. 497–502, 1993.
93. A. Tsagkarakou, M. Navajas, J. Lagnel, and N. Pasteur, "Population structure in the spider miteTetranychus urticae (Acari: Tetranychidae) from Crete based on multiple allozymes," Heredity, vol. 78, no. 1, pp. 84–92, 1997
94. D. F. Greene and E. A. Johnson, "A model of wind disperal of winged or plumed seeds," Ecology, vol. 70, no. 2, pp. 339–347, 1989.

95. A. Okubo and S. A. Levin, "A theoretical framework for data analysis of wind dispersal of seeds and pollen," Ecology, vol. 70, no. 2, pp. 329–338, 1989.
96. R. Beckwitt and R. Chakraborty, "Genetic structure of Pileolaria pseudomilitaris (Polychaeta: Spirorbidae)," Genetics, vol. 96, no. 3, pp. 711–726, 1980.
97. P. Franck, M. Reyes, J. Olivares, and B. Sauphanor, "Genetic architecture in codling moth populations: comparison between microsatellite and insecticide resistance markers," Molecular Ecology, vol. 16, no. 17, pp. 3554–3564, 2007.
98. B. E. Tabashnik and B. A. Croft, "Managing pesticide resistance crop-arthropod complexes: interactions between biological and operational factors," Environmental Entomology, vol. 11, pp. 1137–1144, 1982.
99. M. A. Caprio and B. E. Tabashnik, "Gene flow accelerates local adaptation among finite populations: simulating the evolution of insecticide resistance," Journal of Economic Entomology, vol. 85, pp. 611–620, 1992.
100. J. E. Dunley and B. A. Croft, "Gene flow measured by allozymic analysis in pesticide resistantTyphlodromus pyri occurring within and near apple orchards," Experimental and Applied Acarology, vol. 18, no. 4, pp. 201–211, 1994.
101. K. Goka, "The effect of patch size and persistence of host plants on the development of acaricide resistance in the two-spotted spider mite Tetranychus urticae (Acari: Tetranychidae)," Experimental and Applied Acarology, vol. 23, no. 5, pp. 419–427, 1999.
102. A. Yamamoto, H. Yoneda, R. Hatano, and M. Asada, "Field selection experiments with hexythiazox in the citrus red mite, Panonychus citri (McGregor)," Journal of Pesticide Science, vol. 20, pp. 307–315, 1995.
103. Mh. Osakabe, T. Hirose, and M. Sato, "Discrimination of four Japanese Tetranychus species (Acari: Tetranychidae) using PCR-RFLP of the inter-transcribed spacer region of nuclear ribosomal DNA,"Applied Entomology and Zoology, vol. 37, no. 3, pp. 399–407, 2002.
104. Mh. Osakabe, Y. Kotsubo, R. Tajima, and N. Hinomoto, "Restriction fragment length polymorphism catalogue for molecular identification of Japanese Tetranychus spider mites

(Acari: Tetranychidae),"Journal of Economic Entomology, vol. 101, pp. 1167–1175, 2008.

105. L. A. Meyerson and J. K. Reaser, "Biosecurity: moving toward a comprehensive approach," BioScience, vol. 52, no. 7, pp. 593–600, 2002.
106. Y. Ogawa and Mh. Osakabe, "Development, long-term survival, and the maintenance of fertility inNeoseiulus californicus (Acari: Phytoseiidae) reared on an artificial diet," Experimental and Applied Acarology, vol. 45, no. 3-4, pp. 123–136, 2008.
107. K. Ohtsuka and Mh. Osakabe, "Deleterious effects of UV-B radiation on herbivorous spider mites: they can avoid it by remaining on lower leaf surfaces," Environmental Entomology, vol. 38, no. 3, pp. 920–929, 2009.

Chapter 3

Ex-ante Economic Analysis of Biological Control of Coconut Mite in Benin

Jofrey M. Oleke[1], Victor Manyong[1], Djana Mignouna[1], Aida Isinika[2], Khamaldin Mutabazi[2], Rachid Hanna[3], and Maurice Sabelis[4]

[1]International Institute of Tropical Agriculture, Tanzania

[2]Sokoine University of Agriculture, Tanzania

[3]International Institute of Tropical Agriculture, Cameroon

[4]University of Amsterdam

ABSTRACT

The coconut mite, Aceria guerreronis Keifer, has been identified as one of the pests that pose a threat to the coconut industry in Benin. The study presents the simulation results of the economic benefits of the biological control of coconut mites in Benin using a standard economic surplus model. In the least optimistic scenario, the economy would derive an overall net gain of US$155,213.40. Considered at a discount rate of 12% for the period 2008-2027, net present value was about $207,721, while the internal rates of return or break-even discount rates are high at 13.21%. It is therefore recommended that contemporaneous with the release of natural predators of the coconut mite, plans should be underway for improving research and extension services to coconut farmers in Benin.

INTRODUCTION

Coconut (*Cocos nucifera* Beccari) is the main source of cash income for farmers in the coastal belt of Benin. Oleke, Manyong, Hanna, and Isinika (2010) reported that coconut contributes about 56% of the total head of household cash income among coconut farmers in Benin, equivalent to US$400[1] of the total rural household cash income. Additionally, coconut juice serves as an important supply of fluids and minerals. The oilcake remaining after oil is pressed from copra is used as animal feed (Woodroof, 1970). The coconut shell is used directly as fuel, filler, and extender in the synthesis of plastic and for making household articles. In the coastal belt of Benin, different parts of the coconut plant (truck and leaves) are widely used as building materials, and in recent years the coconut palm wood has been used to make high-quality furniture for the local and export market.

Footnote

[1]Unless otherwise specified, all monetary values in the article are in US dollars.

The coconut, like many plants, is subject to attack by various pests and diseases. The coconut mite, *Aceria guerreronis* Keifer, has been identified among common coconut pests of economic importance. The coconut mite breeds under the perianth (the outer part of the flower consisting of the calyx and corolla, and enclosing the stamen and pistils) of coconuts, where it feeds on the epidermal cells of the meristematic region. Occasionally it feeds on the apical meristem of the coconut seedling. In severe infestations, reduction in nut size and malformation of nuts occur. Consequently, farmers incur economic losses due to the continued presence of pest. The coconut mite is therefore one of the leading exotic pests that pose a threat to the coconut industry in countries such as Benin, Tanzania, India, and Sri Lanka. The coconut mite has proven to be difficult to control. A wide range of chemicals have been employed to control the pest over the past two decades, but the results have been unsatisfactory. Thus, efforts to eradicate it or minimize its damage have been expensive (Pimentel, 2000). Meanwhile, farmers continue to suffer high economic loss (Aquino & Arruda, 1967). Good plant husbandry has been recommended to alleviate the economic impact of the mite on coconut production. In the meantime, research has been directed toward identifying resistant coconut varieties and biological control agents (Pimentel, 2000).

Decreases in coconut yield from mite attack cause loss of income, food insecurity, and poverty for farmers and others in the coconut value chain. A survey was carried out by the Coconut Research Institute (CRI) in Sri Lanka in 2001 to monitor harvested nuts for one year at monthly intervals. The study revealed that the percentage of mite-infested nuts in various Sri Lankan cities was 94.4% in Anuradhapura, 94.5% in Pollonnaruwa, 90.5% in Rajangane, 85.1% in Puttalam, and 69.8% in Kurunegala, with a mean of 77.9% (Peiris, 2006). In Tanzania, reduction in copra yield has been variable from 15-40% (Seguni, 2002). Elsewhere, losses due to extensive premature dropping of fruits have been reported, ranging from 60% in Colombia (Zuluaga & Sánchez, 1971) to 70% in Venezuela (Doreste, 1968) and 10-100% (average 21%) in Tanzania (Seguni, 2002). For the nuts that reach maturity, small-sized nuts cannot be sold at the price of a full size nut, thus reducing the income of the farmer since they fetch lower prices. Peiris (2006) estimated loss of income for coconut growers in Sri Lanka to be 7% from rejected nuts and 43% from small-sized nuts. In Tanzania, loss of farmers' income due to coconut mite is estimated to be about 30-50% (Seguni, 2002). Oleke et al. (2010) similarly reported that coconut production at the household level in Tanzania had declined by about 52% since 2007 largely due to attack by coconut mite.

In order to address the problem of declining production due to coconut mite in Benin, the International Institute of Tropical Agriculture (IITA), in collaboration with University of Amsterdam, has been considering biological control of the coconut mite in order to enhance coconut productivity and profitability in affected countries of Africa. Benin was chosen due to the economic importance of the crop for the livelihood of farmers who depend on it. The identified invasive predators (*Neoseiulus baraki*, *N. paspalivorus*, and *Proctolaelaps bickleyi*) are biologically more efficient in reducing coconut mite populations, as shown from areas that have been explored in Brazil (WOTRO Science for Global Development, 2007).

Economic analyses of biological control programs are a valuable input into the decision-making process for biological control programs. Assessing the impact of agricultural research can assist in providing feedback to the research programs and demonstrating actual benefits of the products of research. The basic criterion is that a biological control program should be considered when the benefits are greater than the costs (Jetter, 2005). However, assessing all costs and benefits to a biological control program presents a challenge due to the different analysis methods. Using the economic surplus approach, this study analyzed the likely costs and benefits of implemeting biological control of the coconut mite in Benin. The specific objective of this study was to estimate the total economic surplus derived from introducing biological control of coconut mite and the distribution of such benefits amongst producers and consumers.

Many studies have indicated the economic benefits of investing in biological control. Two studies—one by Norgaard (1988) and another by Zeddies, Schaab, Neuenschwander, and Herren (2000)—demonstrated the high returns of biological control of cassava mealy bug in Africa. Coulibaly et al. (2004) applied an economic surplus model to assess the economic benefits of the classic control of cassava green mite in West Africa. De Groote, Ajuono, Attignon, Djessou, and Neuenschwander (2003) calculated the economic impact of biological control of water hyacinth in Southern Benin. In Benin, Bokonon-Ganta, De Groote, and Neuenschwander (2002) studied the economic impact of the biological control of mango mealy bug through a survey of mango producers. Alene, Neuenschwander, Manyong, Coulibaly, and Hanna (2006) provided an excellent summary of past economic studies on biological control of major pests in sub-Saharan Africa. While many studies have been carried out on the economic impact of biological control, no specific study has been done on coconut mite in Benin, a gap that this article addresses. The economic surplus approach is a powerful tool to estimate the potential benefits of an agricultural and development program, such as the biological control program of coconut mite.

Norton, Ganoza, and Pomareda (1987) employed an ex-ante economic surplus framework to analyze the potential benefits of agricultural research and extension (R&E) in Peru. The study's projected rates of return to R&E in Peru point to large returns to public investment. Variations in R&E benefits across crops bring to the forefront the issues related to allocation of resources.

Krishna and Qaim (2007) studied the potential impacts of Bt eggplant on the economic surplus and health of people in India. Comprehensive farm survey data were used to project farm-level effects and future adoption rates. Simulations showed that the aggregate economic surplus gains of Bt hybrids could be around $108 million per year. Meanwhile, Bayer, Norton, and Falck-Zepeda (2010) evaluated the economic impact of the regulatory process on four transgenic crops in the Philippines: Bacillus thuringiensis (Bt) rice, ringspot virus-resistant (PRSV) papaya, Bt eggplant, and multiple virus-resistant (MVR) tomato. With Bt eggplant and MVR tomato, the Philippines is modeled as a small closed economy; however, when dealing with Bt rice and PRSV papaya, the country is considered a small open economy. The net benefits (total surplus less total costs) are substantial—$20.5 million for Bt eggplant, $33.5 for MVR tomato, $257.2 for Bt rice, and $240.2 for PRSV papaya.

In their study, Islam and Norton (2007) assessed the potential economic impacts of transgenic salinity- and drought-resistant (SDR) rice in Bangladesh. The ex-ante analysis projected that planting SDR rice over 10 years has a total economic surplus amounting to $302.8 million if no international trade is assumed, of which $184.1 million is producer surplus (PS) and $119.7 million is consumer surplus (CS). The net present value (NPV) of benefits is $215.7 million and the internal rate of return (IRR) is 33.8%. Hareau, Mills, and Norton (2006) conducted an ex-ante evaluation of the economic impact of herbicide-resistant transgenic rice in Uruguay, accounting for multinational market power. They came up with a $1.82 million mean NPV for producers, while $0.55 million will go to the multinational firm.

THEORETICAL FRAMEWORK

In analyzing net benefits from biological control programs, different approaches are used. Such considerations include reversible and irreversible benefits, as well as costs the technology will generate. Kikulwe, Wesseler, and Falck-Zepeda (2008) defined reversible costs and benefits as those that can be reversed after the planting of the crop and do not result in additional ex-post (after stopping production) costs and benefits. On the other hand, irreversible costs and benefits refer to those that will continue to occur even if coconuts are no longer produced or those that cannot be fully reversed. The authors further assert that the reversible and irreversible costs and benefits can be further differentiated into private and non-private costs and benefits. This differentiation is useful for understanding the distribution of costs and benefits between, for example, farmers (private) and society at large (non-private). It is argued by Kikulwe et al. (2008) that the concept of uncertainty and irreversibility has found wide application in decision-making among economists and has a tradition in environmental economics that originated in the early 1970s with papers published by Arrow and Fisher (1974) and Henry (1974); in economics it can even be traced back to Louis Bachelier (1900; Bernstein, 1992). While irreversible costs have their relevance for decision-making, this study utilizes the consumer-surplus approach suggested by Alston, Norton, and Pardey (1995). In the concept of uncertainty and irreversibility, commonly found trigger values are up to 2 and even under such a high trigger value the project would generate a positive value (Wesseler, 2009).

The analysis of economic benefits uses the concepts of supply and demand in partial equilibrium. The economic surplus analysis builds upon the approach first utilized by Griliches (1957) in his pioneering study on hybrid corn in which adoption of a technological innovation fosters a downward shift in the supply curve through reductions in the unit cost of production. From economic theory, a supply function can be derived from production cost data. The supply function is not only affected by price or quantity, but also by any factor that could modify the costs of production and shift the supply curve. Adopting a new technology can therefore have a shifting effect on the supply curve. Economic surplus analysis considers the nature of the market for the commodity and the fact that prices may fall as production changes and supply increases. In conducting economic surplus analysis, market effects stemming from whether the product is widely traded are considered. The consumer surplus models can be used both as closed and open economy models. Closed models are used for the commodities that are produced and consumed within the country. In open economy models—such as that of coconut—commodities are linked to international markets. In Benin, where coconuts are exported to neighbouring countries like Nigeria and Togo, we consider a small open economic model.

With a small open economy, the main assumption in a country such as Benin is that it is "too small" in the world-market trade of coconut products to influence the international price significantly (Figure 1). The world price remains constant and all the benefits of the supply shift accrue to producers. The increased coconut production will shift the supply curve downward from S_0 to S_1, and the domestic demand curve of papayas is assumed to remain unchanged. Consumer surplus thus remains constant, whereas producer surplus increases equal to the area I_0bcI_1.

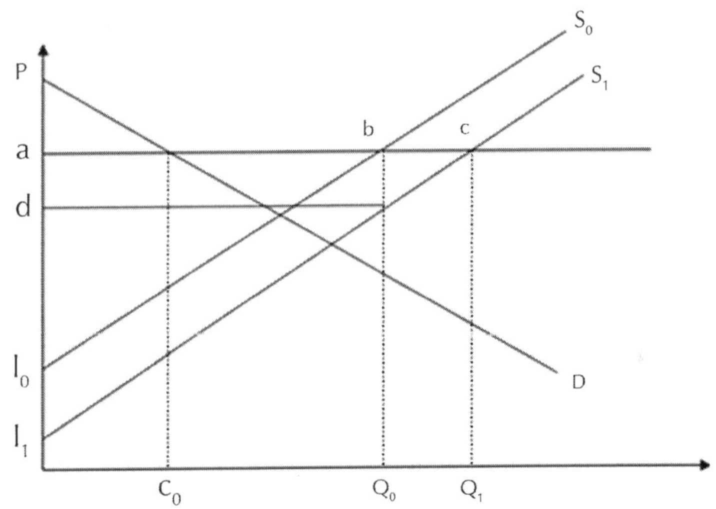

Figure 1: Induced supply shift (small open economy).

ESTIMATING CONSUMER, PRODUCER, AND TOTAL SURPLUS

From the preceding discussion, with the introduction of biological control and the subsequent outward shift of the supply curve, the technology-induced change can be treated as an intercept change (a shift factor *k*) in the supply curve along with the respective quantity supplied and quantity demanded. Following Alston et al. (1995), the absolute relative reduction in price is measured by *Z*. Changes in producer, consumer, and total surplus are estimated by these parameters, *K* and *Z*and are calculated as

$$K = \{[(\mu - \alpha) - P_0 (\beta + \mu)] / \beta\} p_s \text{ and} \quad (1)$$

$$Z = [(P_1 - P_0) / P_1] = - [K\varepsilon_s / (\varepsilon + \eta)] \quad (2)$$

In each year, the downward shift of the supply curve due to technology-induced cost saving from the initial market equilibrium price before the supply shift is P_0, while P_1 represents the new equilibrium price due to increased coconut production;ε_s and η are the supply and demand elasticizes, respectively; μ is the intercept of the demand curve; α and β are the slope of the demand and supply curves, respectively; and p_s is the probability of research success. The changes in producer, consumer, and total surplus in the closed economy are given by

$$\Delta CS = P_0\, Q_0\, Z\, (_{1+0.5}Z\eta)\,, \quad (3)$$

$$\Delta PS = P_0\, Q_0\, (K - Z)\, (_{1+0.5}z\eta), \text{ and} \quad (4)$$

$$\Delta TS = \Delta CS + \Delta PS = P_0\, Q_0\, K\, (_{1+0.5}Z\eta) \quad (5)$$

In case of the small open economy model, the formulae for estimating changes in producer, consumer, and total surplus are given by

$$\Delta SC = 0 \text{ and} \quad (6)$$

$$\Delta PS = \Delta TS = P_w\, Q_0\, K(_{1+0.5}K\varepsilon_s) = P_0\, Q_0\, K\, (_{1+0.5}K\varepsilon_s)\,, \quad (7)$$

where ΔCS is the change in consumer surplus, ΔPS is the change in producer surplus, ΔTS is the change in total surplus, ε_s is the elasticity of supply, η is the elasticity of demand, P_0 is the pre-innovation price,

Q_0 is the pre-innovation quantity produced, P_w is the world price, while Z and K are as previously defined.

METHODOLOGY

The simulation of benefits of biological control of coconut was done using Dynamic Research Evaluation for Management (DREAM) software developed for the International Food Policy Research Institute (IFPRI) by Wood, You, and Baitx (2001). The software calculates and analyzes the benefits derived from technological change, which are measured as economic surplus of producers and consumers. The model also estimates indicators of social gains from investment in research. By introducing into the model the annual flows of investments in research and development, indicators of social gains such as the NPV, IRR, and the cost/benefit (C/B) ratio were estimated.

EMPIRICAL MODELS, DATA, AND ASSUMPTIONS

For this study the baseline levels of production were gathered from an agricultural research station in Semi Podjin, Benin. Coconut yields were estimated at 25,000 tons during the 2008/2009 season. The farm-level price of coconut in Benin was estimates at $45 per ton. Table 1 presents research costs over time for biological control of coconut mites. Table 2 summarizes parameters assumed in each scenario. It is not possible to assess the impact of research without data on the technology it produces. It is assumed that there is no cost savings since most current farmers do not use either pesticides or herbicides. In addition, the costs of biological control are assumed to be the same because there is no anticipated technology fee charged by the governments in the two countries. The discount rate is varied and compared within 12% to 20%.

Table 1: Actual research cost

		Year 1	Year 2	Year 3	Year 4	Total
Total cost ( )1		36 ,430.00	31 ,290.00	25 ,640.00	29 ,930.00	123 ,290.00
Total cost (US$)		52 ,714.21	45 ,276.63	37 ,101.08	43 ,099.20	178 ,191.10
Follow up costs						
Budget line/item	Source	Year 5	Year 6	Year 7	Year 8	Total
Personnel, in-kind	Applicants	12	12	12	12	48
Personnel, in-kind	Collaborators	10	10	10	10	40
Laboratory	University of Amsterdam, IITA, National Research Systems	10	10	10	10	40

Total	-	32	32	32	32	128

Source: WOTRO (2007)

1 Exchange rate €1=$1.44

Table 2: Sources of some of parameters used

Variable	Data/parameter	Source of data
Production quantity, (Q0) (ton)/year	23,121	Semi Podji research station, Benin
Annual production area growth (%)	0.04	Semi Podji research station, Benin
Yield in the start of period of simulation (tons)	25,000	Semi Podji research station, Benin
Price, Po ($/ton)	45	Food and Agricultural Organization of the United Nations (2010)
Export volume, (Qo–Co) (ton)	8,140	Semi Podji research station, Benin
Consumption quantity, Co(ton)/ country (2008)	14,981	Semi Podji research station, Benin
% yield increase, Δ(Y)	50	Interviews with scientists (WOTRO, 2007)
% cost reduction, E(C)	0	-
Elasticity of supply ( )	0.4	Ackah and Appleton (2007)
Elasticity of demand ( )	-0.552	Ackah and Appleton (2007)
Maximum adoption rate (%)	90	Interviews with scientists (WOTRO, 2007)

Lag of R&D returns after first adoption (years)	6	Interviews with scientists (WOTRO, 2007)
Simulation period (years)	20	Interviews with scientists (WOTRO, 2007)

As the technology for biological control of coconut mites using natural predators is still in the development stage, reliable farm-level estimates of changes in yield from its spread could not be obtained. Based on scientist interviews, the probability of success has been set at 90%. Hence, in this study, the full impact of biological control is expected to be a 50% increase of the base scenario in Benin. These estimates are based on responses from scientists from the research program who are involved in collecting predators. The scientist respondents defined the probability of research success as the likelihood that the biological agents will be successfully identified, bred, and released and they in turn successfully suppress the coconut mite. Scientists agreed that the required predator has been identified with complete certainty and multiplied ready for release. It may take time for the biological control agent to completely spread throughout the area infested with coconut mites. Pickett et al. (1996) showed that some agents spread very quickly while others spread at a much slower rate. For simplicity, in this study we assumed that if the program is successful—after introducing the biological control agent—it will take 6 to 10 years to spread throughout the entire mite-infested area, which also marks the end of the program.

If the homogeneity condition holds, the price elasticity for a normal good is in many cases slightly higher than the income elasticity with an opposite sign. The homogeneity condition states that the sum of income elasticity of demand, own-price elasticity of demand, and cross-price elasticities equals 0 (Alston et al., 1995). In Benin, Ackah and Appleton (2007) estimated the elasticity of similar products (palm oil and groundnut) to be -0.552. We use the same base-price elasticity for Benin. Rao (1989) estimated the agricultural supply response to prices in developing countries and found crop-specific acreage elasticities varied from 0 to 0.8 in the short-run and from 0.3 to 1.2 in the long-run for a wide variety of crops. Since no estimates for supply elasticity of coconut and related crops are available for Benin, we use the elasticity estimated in Benin by Ackah and Appleton (2007) for similar commodities—0.4.

SIMULATED EQUILIBRIUM QUANTITIES

Figure 2 shows the simulated supply and demand equilibrium points of the quantities of coconut produced and traded over the period of 20 years. In a small open model (Benin), it is assumed that coconut production will remain constant during the first five years and begin to increase as the effect of predators begins and continues to be felt. While the simulation model shows that production will remain constant (when interest rate 'r' and probability of success 'p' are 12% and 90%, respectively), there may also be some decline in production over the period of five years due to other factors that are not exclusive of coconut-mite attack. The first five years are designed for technology development and, hence, no increase in coconut production is expected. As shown in Figure 2, the benefits from research, in terms of increased production, grow slowly at first. Over the entire period of simulation, analysis of time against yield (tonnes) shows that in Benin yield will increase by about 65.09 tons per year (R^2=89.7%). Meanwhile, production without biological control shows a decrease in yield by 172 tons per year (R^2=93%).

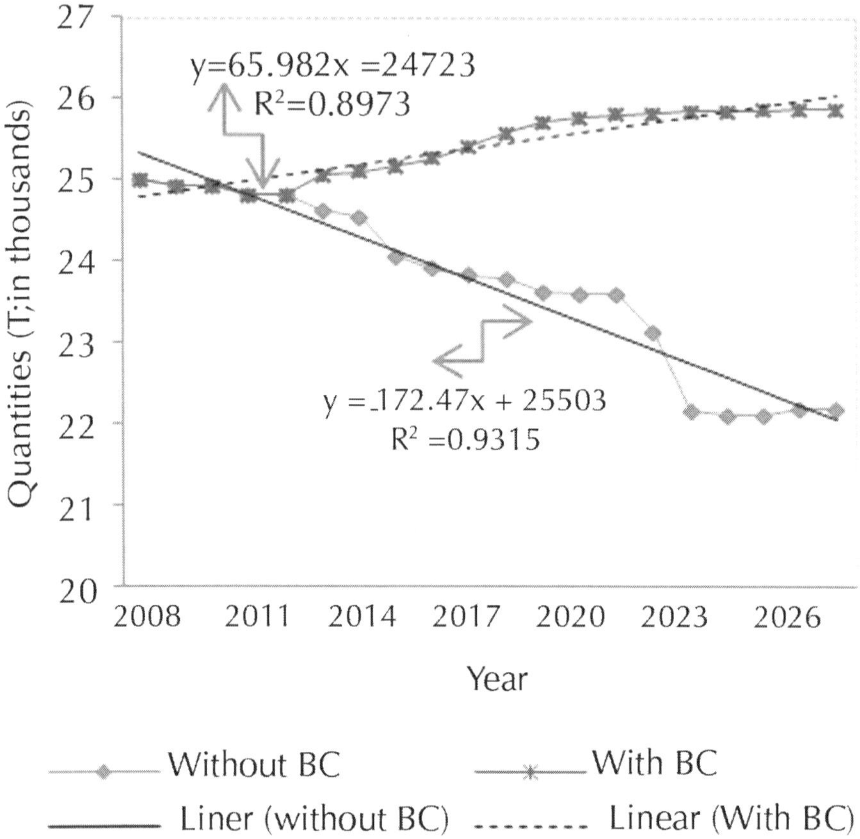

Figure 2: Equilibrium quantities due to biological control of coconut mite.

As shown in Table 3, the research cost is highest before and after the technology is released. At this point, the net benefits (benefits minus costs) do not become positive until spread of biological control is well underway. In the cases shown, the gains increase as the technology continues to spread to a larger part of the affected area. Since Benin is a small exporting economy, the changes in prices due to biological control will not affect the regional markets prevailing in West Africa. Thus, the changes in prices and the corresponding changes in consumer surplus are equal to 0. However, consumers may still benefit from improved quality and continuity of coconut supply.

Table 3: Total benefit

Year	Producer surplus	Consumer surplus	Total benefits	Total cost	Benefits – cost
2008	0	0	0	-52,714.21	-52,714.21
2009	0	0	0	-45,276.63	-45,276.63
2010	0	0	0	-37,101.08	-37,101.08
2011	0	0	0	-43,099.20	-43,099.20
2012	0	0	0	0	0
2013	1,537.6	0	1,537.6	0	1,537.50
2014	3,831.0	0	3,831.0	0	3,831.00
2015	9,136.8	0	9,136.8	0	9,136.70
2016	19,827.5	0	19,827.5	0	19,827.40
2017	36,418.7	0	36,418.7	0	36,418.70
2018	54,022.8	0	54,022.8	0	54,022.70
2019	66,558.6	0	66,558.6	0	66,558.50
2020	73,184.8	0	73,184.8	0	54,022.70
2021	76,153.7	0	76,153.7	0	66,558.70
2022	77,394.4	0	77,394.4	0	73,184.80
2023	77,425.3	0	77,425.3	0	76,153.60
2024	77,426.3	0	77,426.3	0	77,456.30
2025	77,487.3	0	77,487.3	0	77,487.20
2026	77,518.3	0	77,518.3	0	77,518.20
2027	77,549.3	0	77,549.3	0	77,549.20
Discounted totals 1	155,213.4	0	155,213.4	1.01	153,393.60

[1]DREAM software computes total costs and benefits discounted over the entire period of simulation. These totals are not simple additions and averages.

The surplus resulting from agricultural research expenditures was positive and substantial every year in the period considered for simulation. The producers captured the benefit of about $155,213.40 (100%) of total benefits resulting from the investment. These findings are also consistent with those reported by Bayer et al. (2010) that there was no change in consumer surplus due to the use of a small open economy assumption in the study of the cost of compliance with biotechnology regulation on PRSV papaya and Bt rice in the Philippines. Another study by Napasintuwong and Traxler (2009) on an ex-ante impact assessment of GM papaya adoption in Thailand also reported zero change in consumer surplus when a small open model was considered.

SENSITIVITY ANALYSIS

In the analysis, some of the parameters—including the discount rate and the probability of research success—are uncertain because precise information for a future period is lacking. In order to determine the robustness of the simulated returns (net benefits, C/B, and IRR) in view of such uncertainty, these parameters have been varied to check the corresponding changes in the benefits accruing to producers and consumers. The effects of discount-rate changes on returns to investment were estimated by increasing the discount rate from 12% to 20%, commonly used in long-term projects (15-20 years) in developing countries. High discount rates discourage investments with long-term benefits spread over a long period, which include research such as this study.

As shown in Table 4, the total discounted value of economic surplus is $155,213.40 at a discount rate of 12%. If the discounting rate is raised to 15% economic surplus drops to $107,869.10, while the IRR remains the same (13.21%). When a discount rate of 20% is used, it causes further drop in economic surplus to $61,002.80. This means the higher the discount rate, the smaller the gain from the research project. As it is for net benefit, the sensitivity analysis results show that the higher the discount rate, the lower the cost-benefit ratio. For example, when the discount rate is 12%, the C/B ratio is 1.01, whereas a 20% discount rate results in a cost benefit of 0.43.

Table 4: Sensitivity analysis in relation to the discount rate

Discount rate	Producer surplus	Consumer surplus	Total economic surplus	Total cost	(Benefits – cost)	(C/B)	IRR
12%	155,213.3	0	155,213.3	1 ,533 ,393.30	1,819	1.01	13.21
15%	107,869.1	0	107,869.1	148 ,477.36	-4,608	0.73	13.21
18%	76,225.6	0	76,225.6	143 ,961.16	-67,735	0.53	13.21

20%	61,002.8	0	61,002.8	141 ,151.04	-80,148	0.43	13.21

A discount rate of 12% was assumed for computing the IRR, as presented in Table 4. Results show that as the probability of research success increases, the IRR also increases. Furthermore, since one of the criteria to establish the worthiness of the project is IRR, these findings indicate that in order for this project to be beneficial it should be successful by more than 80%.

The NPV of the investment was similarly subjected to sensitivity analysis upon changing the probability of success (Table 5). Again, the probability of success was varied from 20% to 100%. When the probability is assumed to be 20%, NPV reached $34,192. When the probability of research success is increased to 40%, the NPV increases by 100%. Furthermore, if 60% and 80% probability of success were considered, the resulting NPV increased by 50% and 41%, respectively; this implies that as the probability of success increases, the NVP increases at a decreasing rate in both countries. Results in Table 5 show that even if coconut mites are controlled by 40% (which is lower than half of our baseline assumption of 90% probability of research success) the investment will still produce positive net benefit.

Table 5: Net present value for different probability of success

Probability of success	NPV	Change in NPV	% change	IRR (%)
20%	34,192	-	-	-0.05
40%	68,556	34,364	100	5.69
60%	103,090	34,534	50	9.32
80%	145,789	42,699	41	12.06
100%	172,673	26,884	18	14.13

Sensitivity analysis was also done for the NPV in relation to the discount rate as summarized in Table 6. The percentage decline in NPV when the discount rate is reduced from 20% to 12% is about 61%. Decreasing the discount rate from 15% to 12% causes a 30% increase in NPV. Decreasing the discount rate further increases the NPV even more. These results clearly demonstrate that even if the interest rate is set much higher than the base case (12%), the NPV still will be positive, indicating that the investment for biological control of coconut mites is economically viable.

Table 6: NPV and varying discount rate

Discount rate	NPV	Change in NPV (US$)	% change
12%	207,721	-	-
15%	144,349	-63,372	-30
18%	101,996	-42,353	-29
20%	81,622	-20,374	-19

CONCLUSIONS AND IMPLICATIONS

These findings provide consistent evidence to confirm our working hypothesis that biological control of the coconut mite is a viable technology. Sensitivity analysis shows that the results remain robust even when the discount rate is raised up to 20%, which is higher than that of the market rate. Similarly, the investment remains viable even when the probability research is reduced to 80%, at which point the IRR is 12.06%, which is equal to the accepted discount. The results also suggest a positive relationship between program effectiveness (measured in terms of NPV) and net-gains (expressed as producer and consumer surplus). In the least optimistic scenario, the economy in Benin is expected to have a net gain of $155,213.40. On the basis of these results, the study concludes that there is strong evidence that economic returns from investing in this research to control coconut mite are highly significant from zero, and they can withstand wide variation of key variables, including the discount rate and the probability of research success.

Since the natural ecosystem is dynamic and constantly changing, expected benefits can only be maintained if other negative factors that may arise to undermine the positive impacts of this technology will be addressed through well-planned, financed, and evaluated extension services. It is recommended that contemporaneous with the release of natural predators of the coconut mite, plans should be underway for improving research and extension services to coconut farmers in Benin. This technology has proved to be economically viable even when key variables such as the discount rate and the probability of research success are varied over a wide range. It is further recommended that plans should be made to use this technology in other countries such as Ghana, Côte d'Ivoire, and Nigeria, which have suffered considerable loss of coconuts due to coconut mites.

ACKNOWLEDGEMENTS

This study is a part of the WOTRO-Integrated Program. The program deals with classical biological control of the invasive coconut mite in Africa and Asia: integrating biological, molecular, and socio-economic studies in the search and introduction of effective natural enemies. Financial support for this study was provided by the University of Amsterdam. This article benefited from the comments of two anonymous reviewers.

REFERENCES

1. Ackah, C., & Appleton, S. (2007). *Food price changes and consumer welfare in Ghana in the 1990s* (Centre for Research in Economic Development and International Trade [CREDIT] Research Paper No. 07/03). Nottingham, UK: University of Nottingham.
2. Alene, A., Neuenschwander, P., Manyong ,V.M., Coulibaly, O., & Hanna, R., (2006). *The impact of IITA-led biological control of major pests in sub-Saharan African agriculture* (Impact Series). Ibadan, Nigeria: International Institute of Tropical Agriculture (IITA).
3. Alston, J.M., Norton, G.W., & Pardey, P.G. (1995). *Science under scarcity*. Ithaca, NY: Cornell University Press.
4. Aquino, M.L.N., & Arruda, G.P. (1967). *O agente causal da 'necrose do olho do coqueiro' em Pernambuco* [*The causal agent of 'necrosis eye coconut' in Pernambuco*]. Recife, Brazil: Instituto Agronômico de Pernambuco (IPA).

5. Arrow, K., & Fisher, A. (1974). Environmental preservation, uncertainty, and irreversibility. *Quarterly Journal of Economics, 88*, 312-319.
6. Bachelier, L. (1900). *Theory of speculation*. Paris: Gauthier-Villars. Translated by A.J. Boness and reprinted in P. Cootner (1994), *The random character of stock market prices*. Cambridge, MA: MIT Press.
7. Bayer, J.C., Norton, G.W., & Falck-Zepeda, J.B. (2010). Cost of compliance with biotechnology regulation in the Philippines: Implications for developing countries.*AgBioForum, 13*(1), 53-62. Available on the World Wide Web:http://www.agbioforum.org.
8. Bernstein, P.L. (1992). *Capital ideas: The improbable origins of modern Wall Street*. New York: Free Press.
9. Bokonon-Ganta, A., De Groote, H., & Neuenschwander, P. (2002). Socio-economic impact of biological control of mango mealy bug in Benin. *Agriculture, Ecosystems, and Environment, 9*(5), 234-243.
10. Coulibaly, O., Manyong V.M., Yaninek S., Hanna R., Sanginga P., Endamana D. et al. (2004). *Economic impact assessment of classical biological control of cassava green mite in West Africa* (memo). Cotonou, Benin Republic: IITA.
11. De Groote, H., Ajuono, O., Attignon, S., Djessou, R., & Neuenschwander, P. (2003). Economic impact of biological control of water hyacinth in Southern Benin.*Ecological Economics, 45*, 105-117.
12. Doreste, S.E. (1968). El ácaro de la flor del cocotero (*Aceria guerreronis* Keifer) en Venezuela [The coconut flower mite in Venezuela]. *Agronomia Tropical, 18*, 370-386.
13. Food and Agricultural Organization of United Nations (FAO). (2010). *FAOSTAT, agricultural production* [database]. Rome: Author. Available on the World Wide Web: http://faostat.fao.org.
14. Griliches, Z (1957). Hybrid corn: An exploration in the economics of technological change. *Econometrica, 25*(4), 501-522.
15. Hareau, G.G., Mills, B.F., & Norton, G.W. (2006). The potential benefits of herbicide resistant transgenic rice in Uruguay: Lessons for small developing countries. *Food Policy, 31*, 162-179.

16. Henry, C. (1974). Investment decision under uncertainty: The irreversibility effect.*American Economic Review, 64,* 1006-1012.
17. Islam, S.M.F., & Norton, G.W. (2007). Salinity and drought tolerant rice in Bangladesh. In G. Ramasamy, K.N. Selvaraj, G.W. Norton, & K. Vijayaraghavan (Eds.), *Economic and environmental benefits and costs of transgenic crops: Ex-ante assessment* (pp. 72-90). Coimbatore, India: Tamil Nadu Agricultural University.
18. Jetter, K. (2005). Economic framework for decision making in biological control.*Biological Control, 35,* 348-357.
19. Kikulwe, E., Wesseler, J., & Falck-Zepeda, J. (2008). *GM banana in Uganda: Social benefits, costs, and consumer perceptions* (EPTD Discussion Paper 00767). Washington, DC: International Food Policy Research Institute (IFPRI).
20. Krishna, V.V., & Qaim, M. (2007, July 29). *Potential impacts of Bt eggplant on economic surplus and farmers' health in India.* American Agricultural Economics Association annual meeting, Portland, OR.
21. Napasintuwong, O., & Traxler, G. (2009). Ex-ante impact assessment of GM papaya adoption in Thailand. *AgBioForum, 12*(2), 209-217. Available on the World Wide Web: http://www.agbioforum.org.
22. Norgaard, R.B. (1988). The biological control of cassava mealybug in Africa.*American Journal of Agricultural Economics, 72,* 366-371.
23. Norton, G.W., Ganoza, V.G., & Pomareda, C. (1987). Potential benefits of agricultural research and extension in Peru. *American Journal of Agricultural Economics, 69*(2), 247-57.
24. Oleke, J.M., Manyong, V.M., Hanna, R., & Isinika, A.C. (2010, August 23-27).*Stakeholders' analysis of coconut production and perception on coconut mite damage and other constraints in Tanzania.* Presented at the XIII International Congress of Acarology, Recife, State of Pernambuco, Brazil.
25. Peiris, T.S.G. (2006, July 24). *Impact of climate change on coconut industry in Sri Lanka.* Presented at the 3rd International Conference on Climate Impact and Assessment (TICCIA), Cairns, Australia.

26. Pickett, C., Ball, J.C., Casanave, K.C., Klonsky, K.M., Jetter, K.M., Bezark, L.G., et al. (1996). Establishment of the ash whitefly parasitoid Encarsia inaron (Walker) and its economic benefit to ornamental street trees in California. *Biol. Control, 6*, 260-272.
27. Pimentel, D. (2000). Soil erosion and the threat to food security and the environment. *Ecosystem Health, 6*, 221-226.
28. Rao, J.M. (1989). Agricultural supply response: A survey. *Agricultural Economics,3*, 1-22.
29. Seguni, Z. (2002). Incidence, distribution and economic importance of the coconut eriophyid mite, *Aceria guerreronis* Keifer in Tanzanian coconut based cropping systems. In L.C.P. Fernando, G.J. Moraes, & I.R. de Wickramananda (Eds.),*Proceedings of the international workshop on coconut mite* (pp. 54-57). Lunuwila, Sri Lanka: Coconut Research Institute.
30. Wood, S., You, L., & Baitx, W. (2001). *Dynamic research evaluation for management (DREAM), Version 3* [software]. Washington, DC: International Food Policy Research Institute (IFPRI).
31. Woodroof, J.G. (1970). *Coconuts: Production, processing, products*. Ottawa, Ontario, Canada: Aviation Publishers Company.
32. WOTRO Science for Global Development. (2007). *Classical biological control of the invasive coconut mite in Africa and Asia: Integrating biological, molecular and socioeconomic studies in the search and introduction of effective natural enemies*(Integrated Programme—Final Application). The Hague: WOTRO, Netherlands Organisation for Scientific Research (NWO).
33. Zeddies, J., Schaab, R.P., Neuenschwander, P., & Herren, H.R. (2000). Economics of biological control of cassava mealybug in Africa. *Agricultural Economics, 24*, 209-219.
34. Zuluaga, C.I., & Sánchez, P.A. (1971). La roña o escoriación de los frutos del cocotero (*Cocos nucifera L.*) en Colombia [The scabbing or excoriation of the coconut palm fruits in Colombia]. *Acta Agronomica, 21*, 133-139.
35. Wesseler, J. (2009). The Santaniello theorem of irreversible benefits. *AgBioForum,12*(1), 8-13. Available on the World Wide Web: http://www.agbioforum.org.

Chapter 4

Jasmonate-Dependent Plant Defense Restricts Thrips Performance and Preference

Hiroshi Abe[1], Takeshi Shimoda[2], Jun Ohnishi[3], Soichi Kugimiya[4], Mari Narusaka[5], Shigemi Seo[6], Yoshihiro Narusaka[5], Shinya Tsuda[2], and Masatomo Kobayashi[1]

[1]Experimental Plant Division, RIKEN BioResource Center, Tsukuba 305-0074, Japan

[2]National Agricultural Research Center, Tsukuba 305-8666, Japan

[3]National Institute of Vegetable and Tea Science, Tsu 514-2392, Japan

[4]National Institute for Agro-Environmental Sciences, Tsukuba 305-8604, Japan

[5]Research Institute for Biological Sciences, Okayama 716-1241, Japan

[6]National Institute of Agrobiological Sciences, Tsukuba 305-8666, Japan

ABSTRACT

Background

The western flower thrips (*Frankliniella occidentalis* [Pergande]) is one of the most important insect herbivores of cultivated plants. However, no pesticide provides complete control of this species, and insecticide resistance has emerged around the world. We previously reported the important role of jasmonate (JA) in the plant's immediate response to thrips feeding by using an *Arabidopsis*leaf disc system. In this study, as the first step toward practical use of JA in thrips control, we analyzed the effect of JA-regulated *Arabidopsis* defense at the whole plant level on thrips behavior and life cycle at the population level over an extended period. We also studied the effectiveness of JA-regulated plant defense on thrips damage in Chinese cabbage (*Brassica rapa* subsp.*pekinensis*).

Results

Thrips oviposited more on *Arabidopsis* JA-insensitive *coi1-1* mutants than on WT plants, and the population density of the following thrips generation increased on *coi1-1* mutants. Moreover, thrips preferred *coi1-1* mutants more than WT plants. Application of JA to WT plants before thrips attack decreased the thrips population. To analyze these important functions of JA in a brassica crop plant, we analyzed the expression of marker genes for JA response in *B. rapa*. Thrips feeding induced expression of these marker genes and significantly increased the JA content in *B. rapa*. Application of JA to *B. rapa* enhanced plant resistance to thrips, restricted oviposition, and reduced the population density of the following generation.

Conclusion

Our results indicate that the JA-regulated plant defense restricts thrips performance and preference, and plays an important role in the resistance of *Arabidopsis* and *B. rapa* to thrips damage.

BACKGROUND

Insect attack is one of the most important factors retarding plant growth, decreasing crop productivity, and causing other agricultural problems. A constitutive and inducible plant defense response confers immunity to herbivorous insects [1-3]. Analyses at the molecular, metabolic, and physiological levels [2,4] have focused on responses to lepidopteran larvae (caterpillars) and aphids. Many analyses of plant responses to feeding by caterpillars have been conducted [e.g., [5-7]]. Caterpillars harm plants by chewing-type feeding, the best understood of several feeding modes. Although caterpillar feeding and mechanical wounding are physically similar, plants show obvious specific responses to caterpillar feeding [8]. Some of these responses are induced by insect gut and oviposition [9,10]. The sucking-type feeding by aphids and whiteflies is also well understood. However, in contrast to caterpillar feeding, sucking-type feeding rarely causes mechanical damage to the host plant. Rossi et al. [11] reported that the nematode resistance (*R*) gene *Mi-1* of tomato is involved in resistance to the potato aphid. *Mi-1* also confers resistance to whiteflies [12]. Other major classes of insect feeding are also known. Leafminers feed within leaves and stems, forming tunnels (mining-type feeding), and thrips and spider mites feed by piercing and sucking [13,14].

The western flower thrips (*Frankliniella occidentalis* [Pergande]) is one of the most important insect herbivores. This tiny insect tends to occupy narrow crevices within or between plant parts. The emergence worldwide of insecticide resistance among western flower thrips makes them difficult to control [15]. The thrips can also act as a vector of tospoviruses such as tomato spotted wilt virus[16,17]. Damage by western flower thrips is increasing in many countries; in particular, injury in greenhouse production is serious [18-20]. Thus, the development of new methods to control thrips damage by using the molecular mechanisms of plant responses is needed.

Jasmonate (JA) has an important function in plant responses to caterpillars and aphids [2]. Reymond et al. [21] reported that the JA-insensitive *coi1-1* mutant of *Arabidopsis* is less resistant to cabbage butterfly (*Pieris rapae*). Ellis et al. [22] reported that *coi1-1* mutants are less resistant to aphids, but the constitutive JA-signaling mutant *cev1* is more resistant. Our recent study focusing on *Arabidopsis* response to thrips feeding also indicated the important function of JA [23,24], and comparative transcriptome analyses suggested a strong relationship between JA treatment and thrips feeding [23]. Several groups reported that JA-regulated gene expression is induced by spider mites feeding [25,26], which have a similar feeding mode to that of thrips. De Vos et al., using *Arabidopsis* genome arrays [27], also reported the importance of JA for feeding-inducible gene expression by thrips and cabbage butterfly attack. Interestingly, they indicated the existence of common genes in the response to both feeding modes, and genes specific to each feeding mode.

Arabidopsis is a widely studied experimental plant for which many useful genomic resources and much other information are available. However, it is not suitable for analyzing *Arabidopsis*responses to caterpillars, which can quickly eat an entire plant. On the other hand, with the tiny western flower thrips, it is possible to analyze *Arabidopsis* responses to thrips attack over generations.

In this study, we focused on the effect of JA-regulated *Arabidopsis* defense at the whole plant level on thrips behavior and life cycle at the population level. We analyzed the long-term effects of JA-regulated plant defense on thrips oviposition, the population density of the following thrips generation (larvae and pupae), and preference between *Arabidopsis* WT and JA-insensitive *coi1-1*mutant host plants. The results show important effects of the JA-dependent plant defense on both thrips performance and preference. In addition, application of JA to *Arabidopsis* WT plants before thrips attack decreased the thrips population. Expression analyses of marker genes for JA response in Chinese cabbage (*Brassica rapa* subsp. *pekinensis*) suggested the occurrence of a JA-dependent defense against thrips attack in this plant, too. The JA content of *B. rapa* was significantly increased after thrips feeding, and application of JA to plants enhanced their resistance to thrips.

RESULTS

Importance of Jasmonate-Regulated *Arabidopsis* Defense in Resistance to Thrips Attack

We recently reported the role of JA in the short-term response of *Arabidopsis* to thrips feeding on leaf discs over 1 or 2 days [23,24]. To analyze its role in long-term defense at the whole plant level, we compared the feeding damage between whole WT plants and JA-insensitive *coi1-1*mutants [28] inoculated with 20 thrips at 3 weeks. The *coi1-1* mutants had been completely devoured by 4 weeks after inoculation, whereas WT plants were flowering and producing siliques (Fig. 1A). These results suggest the importance of JA-regulated defense in the resistance of*Arabidopsis* to thrips attack.

Figure 1: Function of JA in plant resistance to thrips feeding. Twenty adult females fed on 3-week-old WT plants (left) or *coi1-1*mutants (right). Typical plants after 4 weeks of feeding are shown.

To understand why *coi1-1* mutants showed low resistance to thrips attack, we first analyzed the number of thrips eggs on the WT plants and *coi1-1* mutants to compare the asexual oviposition performance of thrips. *Arabidopsis* rosette leaves were cut into leaf discs with 8-mm diameter. One adult female thrips was put on each disc and allowed to feed and oviposit. Because the females lay in the epidermal or mesophyll cells [29], we stained the eggs with trypan blue to count them. As we reported previously [23], the area of feeding scars on *coi1-1* mutants was greater than that on WT plants (data not shown). The number of eggs on the *coi1-1* discs was double that on the WT discs (Fig. 2A–C). The decreased resistance of these *coi1-1* mutants could explain the increased oviposition rate on these plants.

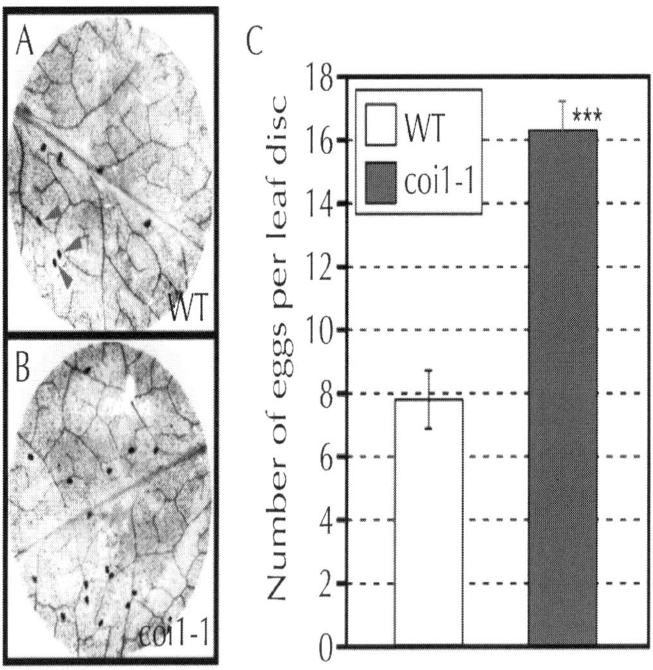

Figure 2: Effect of JA-dependent plant resistance on thrips oviposition on leaf discs. One adult female fed per leaf disc of 3-week-old WT plants (A) or *coi1-1* mutants (B) for 4 days. Eggs oviposited on leaf discs were stained with trypan blue. Photos show typical leaf discs after staining; some eggs are shown by red arrowheads. (C) Number of eggs per leaf disc (mean ± SD) based on more than 10 independent determinations. Asterisks indicate significant difference (Student›s *t*-test), ***$p < 0.001$.

Effect of Jasmonate-Dependent Arabidopsis Defense on Thrips Population

Because the JA-regulated defense affected oviposition, we analyzed its effect on the subsequent generation. We put 20 adult females on WT and *coi1-1* plants and counted adults, larvae, and pupae after 2 weeks. We covered the soil with fine zirconia beads 0.4 mm in diameter to find thrips easily. Thrips fed much more on *coi1-1* mutants than on WT plants (Fig. 3A, B). About 14 of the original adult females remained on *coi1-1* mutants, but only about 2 remained on WT plants (Fig. 3C). Similarly, while more than 190 larvae lived on the *coi1-1* mutants, only about 20 lived on the WT plants (Fig. 3D). We also found 5 times as many pupae on *coi1-1* mutants than on WT plants (Fig. 3E). These results demonstrate that the JA-regulated defense can determine thrips population size.

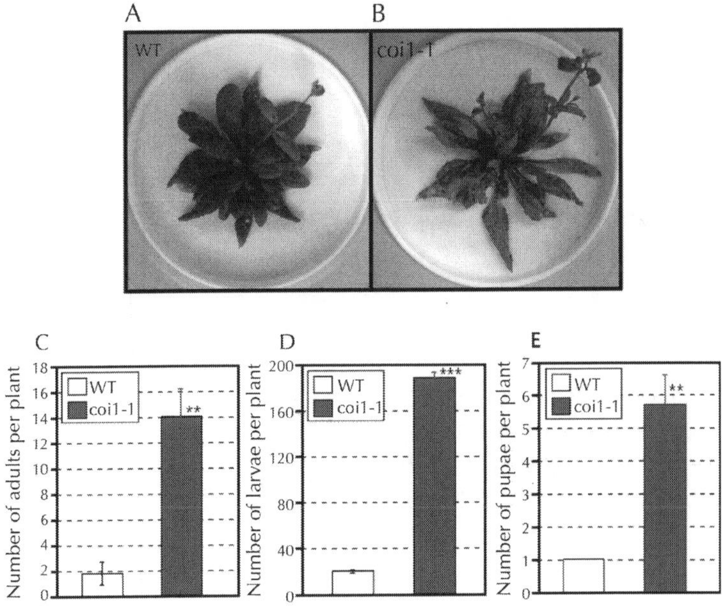

Figure 3: Effect of the JA-dependent plant defense on thrips population. (A, B) Twenty adult females fed on 3-week-old WT plants (A) or *coi1-1* mutants (B) for 2 weeks. (C-E) Number of adults (C), larvae (D), and pupae (E) after 2 weeks; mean ± SD based on five independent determinations. Asterisks indicate significant differences (Student›s *t*-test), $^{**}p < 0.005$, $^{***}p < 0.001$.

Next, we analyzed the effect of JA-regulated plant defense on host plant preference of thrips. We placed 100 adult females halfway between WT and *coi1-1* plants (Fig. 4A) and counted the thrips on each plant after 2 days. The *coi1-1* mutants had many more thrips than the WT plants (Fig. 4A): > 70% versus about 5% (χ^2 test, $\chi^2 = 175.879$, df = 1, $p < 0.001$; Fig. 4B); the remaining thrips roamed the surroundings. These results indicate that the JA-regulated plant defense influences the host plant preference of thrips.

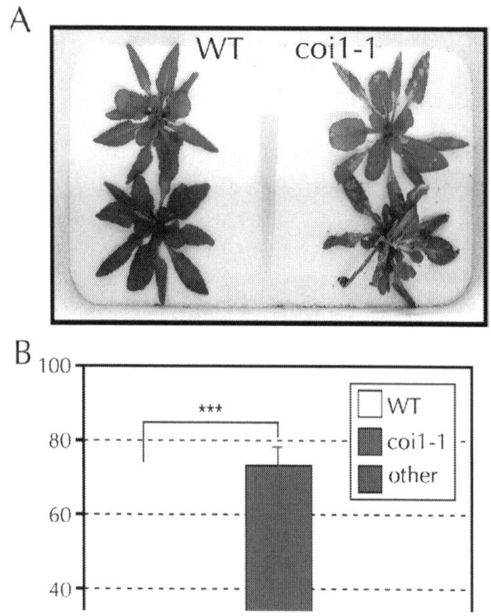

Figure 4: Effect of the JA-dependent plant defense on host plant preference of thrips. (A) Three-week-old WT plants (left) and *coi1-1*mutants (right) were grown at each end of a pot. One hundred adult females were collected in a 1-mL tube and laid between the plants. Photo shows plants after 2 days. (B) Number of adult thrips on each plant after 2 days. Mean ± SD based on five independent determinations. Asterisk indicates a significant difference between the two plants, χ^2 test, ***$p < 0.001$. The remaining thrips roamed the surroundings were excluded from the statistical analysis.

We next analyzed the effect of JA treatment on *Arabidopsis* resistance to thrips attack. JA-treated plants had half as many eggs as untreated plants (Fig. 5A). The numbers of adults and larvae showed a similar contrast (Fig. 5B, C). Together with the results from the *coi1-1* mutants, these results indicate that the JA-dependent defense response in *Arabidopsis* plays an important role in resisting thrips.

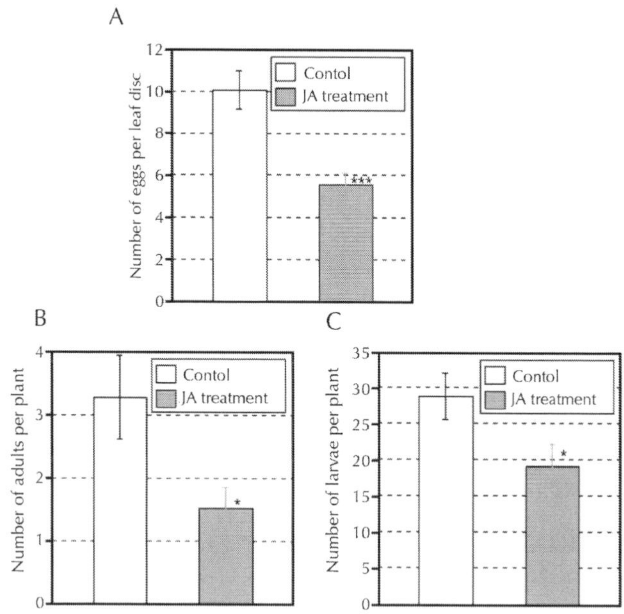

Figure 5: Effect of JA-induced *Arabidopsis* defense response on thrips population. Twenty adult females fed on 3-week-old WT plants. Either 50 μM JA or water (control) was applied 2 days before thrips were introduced. After 2 weeks, eggs (A), adults (B), and larvae (C) were counted. Mean ± SD based on five independent determinations. Asterisks indicate significant differences (Student›s *t*-test), *$p < 0.05$, ***$p < 0.001$.

Jasmonate-Dependent Plant Resistance to Thrips in *B. Rapa*

To search for JA-dependent resistance to thrips in a brassica crop, we analyzed the function of JA in *B. rapa*, one of the most important brassica crops in the world, especially in Asia. A search of the *B. rapa* EST database (National Center for Biotechnology Information) revealed putative counterparts of *Arabidopsis* JA-inducible marker genes. We analyzed the expression of marker genes of the JA pathway corresponding to *AtVSP2* and *AtLOX2*, and genes corresponding to *allene oxide synthase* (*AtAOS*) and *allene oxide cyclase 2* (*AtAOC2*), both of which encode enzymes that catalyze JA biosynthesis as shown by previous reports (Fig. 6E) [30,31]. Expression of the brassica counterparts, *BrVSP2*, *BrLOX2*, *BrAOS*, and *BrAOC2*, was induced by thrips feeding (Fig. 6A–D). In addition, the JA content of *B. rapa* plants infested by thrips was significantly higher than that of control plants (one-way ANOVA, $F = 13.938$, df = 2, $p < 0.01$; Fig. 6F). These data suggest the involvement of JA in the response to thrips feeding in *B. rapa* also.

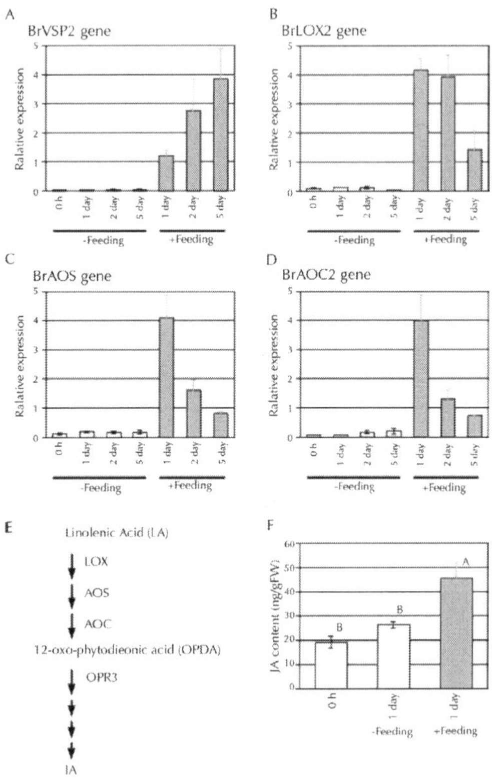

Figure 6: Involvement of JA signaling in *B. rapa* response to thrips feeding. (A-D) Expression of marker genes for JA response in *B. rapa*was induced by thrips feeding. *BrVSP2* (A), *BrLOX2* (B), *BrAOS* (C), and*BrAOC2* (D) are brassica counterparts of *Arabidopsis* marker genes of the JA pathway and JA biosynthesis. Twenty-five adult females fed on five 2-week-old plants per pot. After 0, 1, 2, and 5 days, total RNA was prepared from the plants with (+Feeding) or without (-Feeding) thrips, and first-strand cDNA was synthesized for PCR analysis. The expression level of each gene was normalized to the expression

of *BrACT2* (control). Mean ± SD based on three replications. (E) Proposed model of the biosynthesis of JA in *Arabidopsis*. (F) Effect of thrips feeding on the biosynthesis of JA in *B. rapa*. Ten adult females fed on a 2-week-old plant (+Feeding). A control plant was kept without thrips (-Feeding). At the beginning of the experiment (0 h) and after 1 day from the start of feeding, 1 g of plant tissue was sampled for measurement of endogenous JA (JA + methyl JA). Means ± SD of three independent measurements. Different letters indicate statistically significant differences between treatments (Tukey-Kramer HSD test; $p < 0.05$).

To confirm the functional role of JA in plant resistance to thrips attack in *B. rapa*, we analyzed the effect of JA treatment on thrips feeding. Injury from thrips attack was lower in plants treated with JA than in untreated plants (Fig. 7A–D), by a factor of about 15 (Fig. 7E). These results indicate that the JA-dependent plant defense against thrips is conserved in *B. rapa*.

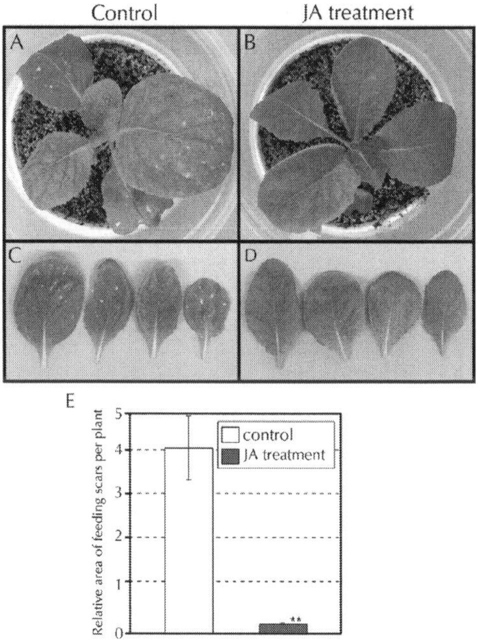

Figure 7: Effect of JA application on plant resistance to thrips. Twenty adult females fed on 2-week-old *B. rapa* plants for 10 days. Water (control; A, C) or 50 µM JA (B, D) was applied 1 day before thrips were introduced. (E) Mean ± SD of area of feeding scars based on more than 10 independent determinations. Asterisk indicates significant difference (Student›s *t*-test), $^{**}p < 0.005$.

We further analyzed JA's effect on thrips oviposition. Rosette leaves of *B. rapa* were cut into leaf discs with 8-mm diameter. One adult female thrips was put on each leaf disc and allowed to feed and oviposit for 4 days. Application of JA dose-dependently decreased the number of eggs (one-way ANOVA, $F = 10.367$, df = 4, $p < 0.001$; Fig. 8A). Finally, we analyzed the effect of JA on the next generation. JA treatment of plants restrained the thrips population very effectively (Fig. 8B, C). These results clearly indicate the important role of JA in resistance to thrips attack in *B. rapa*also.

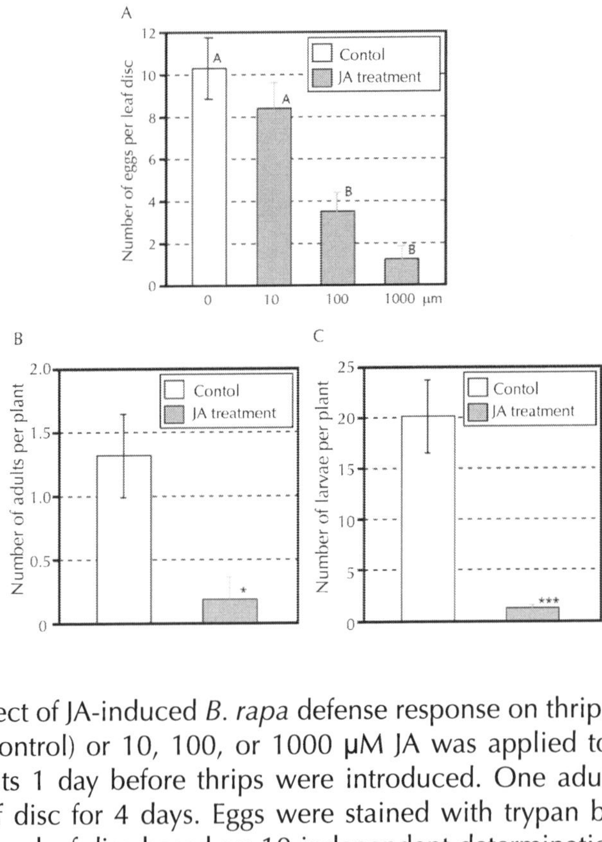

Figure 8: Effect of JA-induced *B. rapa* defense response on thrips population. (A) Water (control) or 10, 100, or 1000 µM JA was applied to 2-week-old *B. rapa* plants 1 day before thrips were introduced. One adult female fed on each leaf disc for 4 days. Eggs were stained with trypan blue. Mean ± SD of eggs per leaf disc based on 10 independent determinations. Different letters indicate statistically significant differences between treatments (Tukey-Kramer HSD test; $p < 0.05$). (B, C) Twenty adult females fed on 2-week-old WT plants for 2 weeks. Water (control) or 50 µM JA was applied 1 day before thrips were introduced. After 2 weeks, adults (B) and larvae (C) were counted. Mean ± SD based on five independent determinations. Asterisks indicate significant differences (Student›s *t*-test), *$p < 0.05$, ***$p < 0.001$.

DISCUSSION

The phytohormone JA regulates part of a plant's basal defense system. Numerous studies have examined the functions of JA in plant responses to pathogen attack, mechanical wounding, UV irradiation, ozone exposure, osmotic stress [32,33], and insect feeding [34,35]. The JAZ (jasmonate ZIM-domain) family of repressors was identified in *Arabidopsis* as a negative regulator of JA signaling [36-38]. JAZ interacts with COI1 protein, degrades, and so induces JA-responsive gene expression. Overexpression of a modified form of JAZ1 significantly decreased plant resistance to the beet armyworm (*Spodoptera exigua*) [39]. Resistance by *coi1-1* mutants to cabbage butterfly caterpillar (*Pieris rapae*) was similarly decreased [21].

However, these analyses focused on plant responses to lepidopteran larvae. Because caterpillars quickly devour *Arabidopsis* plants and change to butterflies or moths, which fly away, it is difficult to analyze the *Arabidopsis* response and insect performance over generations on the one*Arabidopsis* plant. For these reasons, we used thrips. We found differences in symptoms between WT plants and JA-insensitive *coi1-1* mutants: thrips had demolished *coi1-1* mutants after 4 weeks, yet WT plants had flowers and siliques (Fig. 1A). As it seemed unlikely that only 20 adult thrips could kill a plant in 4 weeks, we also studied the effect of a JA-dependent *Arabidopsis* defense on oviposition. The number of eggs on *coi1-1* was about double that on the WT (Fig. 2A–C). As we described previously [23], the area of feeding scars in *coi1-1* was much greater than that on WT plants (data not shown). The greater number of eggs on *coi1-1* might result from the better performance of adult thrips. Alternatively, a difference in plant metabolites between WT and *coi1-1*might influence oviposition. Annadana et al. [40] reported that cysteine protease inhibitors restrict oviposition by western flower thrips. Wounding and JA induce many genes encoding cysteine protease inhibitors [41], including *Arabidopsis* cystatin-1 (AtCYS1) [42]. Cysteine protease inhibitors could explain the difference in thrips oviposition between WT and *coi1-1* plants.

Next, we analyzed the effect of JA-regulated plant defense on the population density of the following generation of thrips. Surprisingly, the population increased around 10-fold after 2 weeks on the *coi1-1* mutants, but changed little on the WT plants (Fig. 3A–E). Most of the thrips on *coi1-1*were larvae. We found some dead larvae on the WT plants but none on *coi1-1* (data not shown). These results indicate that the JA-dependent plant defense in WT plants reduces the survival of thrips larvae. We found about 7 times as many adult thrips on *coi1-1* as on the WT, which indicates that thrips can survive longer on *coi1-1*. We attribute the much greater population of thrips on*coi1-1* to this increased longevity and the greater egg production on *coi1-1* mutants, and the higher mortality of larvae on the WT plants. Analysis of the hatching rate of eggs could also help explain the increased population on *coi1-1*. Barth et al. [43] reported that a double knock-out mutant of *Arabidopsis* lacking two major genes for myrosinase (*tgg1, tgg2*), which degrades glucosinolates to toxins such as isothiocyanates, showed decreased resistance to the cabbage looper (*Trichoplusia ni*) and tobacco hornworm (*Manduca sexta*). Sasaki-Sekimoto et al. [33] reported that JA regulates glucosinolate biosynthesis. Recently, Shroff et al. [44] showed that the preferential allocation of glucosinolates to the periphery of leaves may play a key role in the defense of leaves by creating a barrier to chewing herbivores, which frequently approach leaves from the edge. Several other compounds protect plants against insect pests. Konno et al. [45] reported that cysteine proteases such as papain, ficin, and bromelain showed toxicity to two notorious pests, cabbage armyworm (*Mamestra brassicae*) and cotton leafworm (*Spodoptera litura*). They later reported that sugar-mimic alkaloids were toxic to cabbage armyworm [46]. Further analyses will help to explain which kinds of compounds, regulated by JA, reduce thrips performance.

The choice test showed that *coi1-1* mutants attracted 14 times as many thrips as did WT plants (Fig. 4A, B). As a result, *coi1-1* mutants suffered more damage. Aharoni et al. [47] reported that overexpression of a gene for a dual linalool/nerolidol synthase (*FaNES1*) in *Arabidopsis*, which produces those two terpenes, enhances avoidance by green peach aphids (*Myzus persicae*). Interestingly, these *FaNES1*-overexpressing plants also attracted carnivorous predatory mites (*Phytoseiulus persimilis*) [48]. JA-deficient *spr2* tomato plants emit less herbivory-induced volatiles and attract more tobacco hornworm and tobacco whitefly (*Bemisia tabaci*) for oviposition [49]. In addition to the volatile components, many other plant metabolites such as nutrient factors and toxic compounds are reported as stimulants or deterrents of host plant preference [50]. These metabolic components may explain the higher preference of the thrips for *coi1-1* mutants or higher avoidance of WT plants.

The western flower thrips is one of the most serious insect herbivores in the world. It is also a virus vector. Because of its thigmokinetic behavior and the emergence of insecticide resistance, it is difficult to control with insecticides [15]. Therefore, new control methods are urgently needed. Application of JA to WT *Arabidopsis* plants before thrips damage decreased the thrips population (Fig. 5A–C). We previously reported that thrips feeding induced in *Arabidopsis* expression of*AtVSP2* and *AtLOX2* (marker genes of the JA pathway) and *AtAOS1* and *AtAOC2* (encoding allene oxide synthase and allene oxide cyclase), which catalyze JA biosynthesis in *Arabidopsis* [23]. Here, the expression of their counterparts in *B. rapa* was also induced by thrips feeding (Fig. 6A–D), as was the JA content (Fig. 6F), as reported previously in *Arabidopsis* [23]. These results indicate that the JA-dependent defense system is conserved between *Arabidopsis* and *B. rapa*. Interestingly, JA application also greatly decreased the amount of feeding scars in *B. rapa* (Fig. 7A–E), and decreased egg production and thrips population size (Fig. 8A–C). The effect of JA application was much higher in *B. rapa* than in *Arabidopsis*, but the biological significance of this difference is unclear. Several groups have combined JA-mediated transcriptome analyses with metabolomics data [33,51]. Further comparative analyses between *B. rapa* and *Arabidopsis* using these approaches are needed to explain the differences in plant resistance. The genome of *B. rapa* is being sequenced http://brassica.bbsrc.ac.uk/. In the near future, *Brassica* 'omics' analyses using genome information will be available. Comparative expression analyses between *B. rapa* and *Arabidopsis* suggested the existence of similar and specific responses to pathogen infection in these species [52].

Jasmonate application to *Nicotiana sylvestris* plants decreased plant biomass [53]. Overexpression of *AtJMT* in *Arabidopsis* plants, which leads to elevated JA level [54], decreased the flower number and total seed weight significantly. Importantly, Thaler et al. showed that although application of JA in tomato fields successively decreased naturally occurring thrips, spray application at low concentration (0.5 mM) decreased neither plant biomass nor fruit production [55]. However, the effect of low JA concentration on thrips control is lower than that of high JA concentration (1.5 mM). JA application incurs costs for plant fitness, and also activates plant defense, which must be balanced for optimum production. The screening of the specific compounds to regulate plant defense to insect attack will be a promising approach.

CONCLUSIONS

In this study, as the first step toward practical use of JA in thrips control, we analyzed the effect of JA-regulated *Arabidopsis* defense at the whole plant level on thrips behavior and life cycle at the population level. Our results indicate that JA-regulated *Arabidopsis* defense restricts both thrips performance and preference. Thrips performance was evaluated from oviposition and the population density of the following generation. The effect of JA-regulated defense on thrips population density was considerable. This was due to the effects on thrips longevity, egg production, and mortality of larvae. Fully understanding the plant defense against thrips attack will require determination of the actual plant metabolites that restrict thrips performance and preference.

In *B. rapa* also, induction of expression of marker genes for the JA pathway and increased JA content after thrips damage support the occurrence of a JA-dependent defense against thrips attack. JA application to *B. rapa* greatly decreased feeding damage on account of decreased egg production and thrips population density. The existence of diverse targets of JA-regulated plant defense indicates that JA concurrently regulates multiple responses involved in plant resistance to thrips damage. JA-regulated plant defense could be a good target for practical applications to control thrips.

METHODS

Plant Materials and Cultivation

Wild-type (ecotype Col-0) *Arabidopsis* plants and the JA signaling and biosynthesis mutant *coi1-1*were grown in soil as described previously [56]. Briefly, seeds were sown on sterile soil in pots, moistened, and held at 4°C for 7 days in the dark to synchronize germination. The pots were then transferred to 22°C with a long-day photoperiod (16 h light/8 h dark). Plants at the four-leaf stage were transferred individually to pots and grown to the rosette stage. Chinese cabbage (*B. rapa*subsp. *pekinensis* cv. Kyoto No. 3, Takii Seed Co. Ltd., Kyoto, Japan) plants were grown similarly.

Identification of Coi1-1 Plants

Homozygous *coi1-1* plants were selected according to PCR amplification of a sequence of the*Arabidopsis COI1* gene followed by digestion with *Bsm*I (TOYOBO, Osaka, Japan). Within the amplified PCR product, the *Bsm*I restriction site is present only in the *coi1-1* mutant. Primers were as follows: forward, 5′-GGAAACAGGAGCCCGAGATC-3′; reverse, 5′-TGGATGTTTCTCGGAGCAGC-3′.

Thrips Attack

Laboratory colonies of *Frankliniella occidentalis* were maintained in a closed environmental chamber, as described previously [57]. The assay used female adults 14–21 days after emergence from the pupal stage. The adults were starved for 2 to 3 h before feeding on test plants. Twenty adult females were allowed to feed on each whole plant in a cylindrical acryl chamber with air ventilation windows covered with a fine mesh.

Jasmonate Treatment

Pots holding 3-week-old *Arabidopsis* plants or 2-week-old *B. rapa* plants grown in soil were transferred into a cylindrical acryl chamber containing 100 μM JA solution. Other experiments to count the number of eggs on *B. rapa* leaf discs used 10, 100, or 1000 μM JA solution. JA treatment was carried out for 2 days before the beginning of thrips attack.

Counting of Thrips Eggs

Leaf discs with 8-mm diameter were cut with a biopsy punch (Kay Industries, Oyana, Japan). The discs were floated on 1.5 mL of distilled water in wells of a white 1.5-mL sample tube stand (Assist, Tokyo, Japan). A single adult female that had been starved for 2 to 3 h was placed on each leaf disc. The sample tube stand was covered with ABI Prism Optical Adhesive Cover (Applied Biosystems, Foster City, CA, USA), and a few tiny holes for air were made with a 27-G fine injection needle. Thrips were allowed to feed and oviposit for 4 days at 22°C. Eggs were stained with trypan blue as described previously [58].

Counting of the Thrips Population

Three-week-old *Arabidopsis* plants or 10-day-old *B. rapa* plants grown in soil covered with fine zirconia beads (Nikkato Co., Osaka, Japan; 0.4 mm in diameter to make it easy to find the thrips) were placed in a cylindrical acryl chamber as above. Twenty adult females were put on each plant. After 2 weeks, the adults, larvae, and pupae were counted.

Choice Assay

Three-week-old WT and *coi1-1* plants grown in soil covered with fine zirconia beads in a white pot (255 × 145 × 120 mm; Appleware, Osaka, Japan) were used for a choice assay in a cylindrical acryl chamber as above. Each pot held four plants (two of each type) separated by 150 mm. One hundred adult females were deposited halfway between the plants and allowed to move freely. After 2 days, the thrips on each plant were counted.

Quantitative Reverse Transcription PCR

Twenty-five female adult thrips fed on five 2-week-old *B. rapa* plants at the rosette stage for 1, 2, or 5 days in a closed container with air vents. Experiments were repeated twice. After feeding, the plants were frozen in liquid nitrogen. Total RNA (2 μg) isolated with Trizol reagent (Invitrogen, Carlsbad, CA, USA) and an RNeasy MinElute Cleanup Kit (Qiagen, Valencia, CA, USA) was treated with RNase-free DNase (Takara) to eliminate genomic DNA. First-strand cDNA was synthesized with random oligo-hexamers and Superscript III reverse transcriptase according to the manufacturer›s instructions (Invitrogen). Quantitative real-time PCR was carried out with Power SYBR Green PCR Master Mix (Applied Biosystems) using the first-strand cDNA as a template on a sequence detector (ABI Prism

7900HT, Applied Biosystems). Expression of *BrACT2* was used for normalization. Nucleotide sequences of the gene-specific primers were as follows: *BrVSP2* (forward, 5′-GACTCCAAAACGGTGTGCAAA-3′; reverse, 5′-AGGGTCTCGTCAAGGTCAAAGA-3′); *BrLOX2* (5′-TCCCCACTTCCGCTACACC-3′; 5′-AATACTTTCCGGGCCAGAAAC-3′); *BrAOS* (5′-GATCTCCCCATCCGAACCAT-3′; 5′-AACTCCTCGGGTTTTTGCTTG-3′); *BrAOC2* (5′-GCCGGTCTCTGTGTCTTGATC-3′; 5′-ACGGACAGGTGGCCATAGTC-3′); and *BrACT2* (5′-ACCCAAAGGCCAACAGAGAG-3′; 5′-CTGGCGTAAAGGGAGAGAACA-3′).

Jasmonate Quantification

JA and its methyl ester were quantified as described previously [47], except that an HP6890 gas chromatograph fitted to a quadrupole mass spectrometer (Hewlett-Packard, Wilmington, DE, USA) was used. Approximately 1 g of each *B. rapa* plant with or without thrips feeding was used for quantification. Three independent samples were analyzed.

Measurement of the Area of Feeding Scars

The area of thrips feeding scars on the surface of each *B. rapa* leaf was measured using WinROOF software, version 5.8.1 (Mitani Corporation, Tokyo, Japan), on digitized images taken under a VHX-200 digital microscope (Keyence, Osaka, Japan).

Statistics

The results of thrips oviposition, population density and feeding activity were respectively subjected to Student's *t*-test or analysis of variance (one-way ANOVA) followed by Tukey-Kramer HSD test. The result from choice assay was subjected to a χ^2 test; the null hypothesis was that thrips exhibited a 50:50 distribution over WT and *coi1-1* plants. These analyses were performed the JMP software, ver. 5.1 (SAS Institute, Inc., Cary, NC, USA).

Accession Numbers

The GenBank accession numbers for the genes mentioned in this article are as follows: *BrVSP2*(EX101964), *BrLOX2* (EX100417), *BrACT2* (EX137335), *BrAOS* (EX104579), *BrAOC2* (EX125486).

AUTHORS' CONTRIBUTIONS

HA planned the study and carried out the feeding assay, thrips performance and preference analyses, and wrote the manuscript. TS and SK performed thrips performance and preference analyses and participated in final writing of the manuscript. JO did the measurement of the area of feeding scars. MN and YN did the gene expression analyses of Brassica. SS performed JA quantification. ST and MK planned the study and participated in its coordination and final writing of the manuscript. All authors read and approved the final manuscript.

ACKNOWLEDGEMENTS

We thank F. Mori, S. Kawamura, and I. Sasaki of RIKEN BRC and S. Nagai and Y. Matsumura of the National Agricultural Research Center for their excellent technical assistance. We also thank Dr. M. Watanabe of NIAS for his kind support and helpful discussion. This work was supported by a Grant-in-Aid for Scientific Research for a "Young Scientist (B)" from the Ministry of Education, Culture, Sports, Science and Technology to HA, and by a grant from the 2004 Industrial Technology Research Grant Program of the New Energy and Industrial Technology Development Organization of Japan to YN and HA.

REFERENCES

1. Kessler A, Baldwin IT: Plant responses to insect herbivory: the emerging molecular analysis. *Annu Rev Plant Biol* 2002, 53:299-328.
2. Howe GA, Jander G: Plant immunity to insect herbivores. *Annu Rev Plant Biol* 2008, 59:41-66.

3. Howe GA, Schaller A: Direct defenses in plants and their induction by wounding and insect herbivores. In *Induced Plant Resistance to Herbivory*. Edited by Schaller A. New York, Springer; 2008:7-29.
4. Van Poecke RMP: Arabidopsis-insect interactions.[http://www.aspb.org/publications/arabidopsis/] *The Arabidopsis Book* Rockville, MD: American Society of Plant Biologists; 2008.
5. Kahl J, Siemens DH, Aerts RJ, Gabler R, Kuhnemann F, Preston CA, Baldwin IT: Herbivore-induced ethylene suppresses a direct defense but not a putative indirect defense against an adapted herbivore. *Planta* 2000, 210:336-342.
6. Winz RA, Baldwin IT: Molecular interactions between the specialist herbivore *Manduca sexta* (Lepidoptera, Sphingidae) and its natural host *Nicotiana attenuata*. IV. Insect-induced ethylene reduces jasmonate-induced nicotine accumulation by regulating putrescine N-methyltransferase transcripts. *Plant Physiol* 2001, 125:2189-2202.
7. Li C, Liu G, Xu C, Lee GI, Bauer P, Ling HQ, Ganal MW, Howe GA: The tomato *suppressor of prosystemin-mediated responses2* gene encodes a fatty acid desaturase required for the biosynthesis of jasmonic acid and the production of a systemic wound signal for defense gene expression. *Plant Cell* 2003, 15:1646-1661.
8. Reymond P, Weber H, Damond M, Farmer EE: Differential gene expression in response to mechanical wounding and insect feeding in Arabidopsis. *Plant Cell* 2000, 12:707-719.
9. Korth KL: Profiling the response of plants to herbivorous insects. *Genome Biol* 2003, 4:221.1-221.4.
10. Little D, Gouthier-Darimont C, Bressow F, Reymond P: Oviposition by pierid butterflies triggers defense responses in Arabidopsis. *Plant Physiol* 2007, 143:784-800.
11. Rossi M, Goggin FL, Milligan SB, Kaloshian I, Ullman DE, Williamson VM: The nematode resistance gene *Mi* of tomato confers resistance against the potato aphid. *Proc Natl Acad Sci* 1998, 95:9750-9754.
12. Nombela G, Williamson VM, Muniz M: The root-knot nematode resistance gene *Mi-1.2* of tomato is responsible for resistance against the whitefly *Bemisia tabaci*. *Mol Plant Microbe Interact* 2003, 16:645-649.

13. Geest LPS: Aspect of physiology. In *Spider mite: their biology, natural biology, natural enemies and control*. Edited by Helle W, Sabelis MW. Amsterdam, Elsevier; 1985:171-184.
14. Parrella MP: IPM: approaches and prospects. In *Thrips Biology and Management*. Edited by Parker BL, Skinner M, Lewis T. New York, Plenum; 1995:357-364.
15. Jensen SE: Insecticide resistance in the western flower thrips, *Frankliniella occidentalis*. *Integr Pest Manag Rev* 2000, 5:131-146.
16. Wijkamp I, van Lent J, Kormelink R, Goldbach R, Peter D: Multiplication of tomato spotted wilt virus in western flower thrips. In *Thrips Biology and Management*. Edited by Parker BL, Skinner M, Lewis T. New York, Plenum; 1995:357-364.
17. Whitfield AE, Ullman DE, German TL: Tospovirus-thrips interactions. *Annu Rev Phytopathol* 2005, 43:459-489.
18. Lewis T: Pest thrips in perspective. In *Thrips as Crop Pests*. Edited by Lewis T. Cambridge, Cambridge University Press; 1997:1-14.
19. Van Driesche Lyon S, Stanek EJ III, Xu B, Nunn C: Evaluation of efficacy of *Neoseiulus cucumeris* for control of western flower thrips in spring bedding crops. *Biological Control* 2006, 36:203-215.
20. Reitz SR: Biology and Ecology of the Western Flower Thrips (Thysanoptera: Thripidae): The Making of a Pest. *Florida Entomologist* 2009, 92:7-13.
21. Reymond P, Bodenhausen N, Van Poecke RM, Krishnamurthy V, Dicke M, Farmer EE: A conserved transcript pattern in response to a specialist and a generalist herbivore. *Plant Cell* 2004, 16:3132-3147.
22. Ellis C, Karafyllidis I, Turner JG: Constitutive activation of jasmonate signaling in an Arabidopsis mutant correlates with enhanced resistance to *Erysiphe cichoracearum,Pseudomonas syringae*, and *Myzus persicae*. *Mol Plant Microbe Interact* 2002, 15:1025-1030.
23. Abe H, Ohnishi J, Narusaka M, Seo S, Narusaka Y, Tsuda S, Kobayashi M: Function of jasmonate in response and tolerance of Arabidopsis to thrips feeding. *Plant Cell Physiol* 2008, 49:68-80.

24. Abe H, Ohnishi J, Narusaka M, Seo S, Narusaka Y, Tsuda S, Kobayashi M: Arabidopsis-thrips system for analysis of plant response to insect feeding. *Plant Signaling & Behav* 2008, 3:1-2.

25. Arimura G, Ozawa R, Shimoda T, Nishioka T, Boland W, Takabayashi J: Herbivory-induced volatiles elicit defence genes in lima bean leaves. *Nature* 2000, 406:512-515

26. Zheng SJ, van Dijk JP, Bruinsma M, Dicke M: Sensitivity and speed of induced defense of cabbage (Brassica oleracea L.): Dynamics of BoLOX expression patterns during insect and pathogen attack. *Molecular Plant-Microbe Interactions* 2007, 20:1332-1345

27. De Vos M, Van Oosten VR, Van Poecke RMP, Van Pelt JA, Pozo MJ, Mueller MJ, Buchala AJ, Metraux JP, Van Loon LC, Dicke M, Pieterse CMJ: Signal signature and transcriptome changes of Arabidopsis during pathogen and insect attack. *Molecular Plant-Microbe Interactions* 2005, 18:923-937.

28. Xie DX, Feys BF, James S, Nieto-Rostro M, Turner JG: *COI1* : an Arabidopsis gene required for jasmonate-regulated defense and fertility. *Science* 1998, 280:1091-1094.

29. Childers CC, Achor DS: Thrips feeding and oviposition injuries to economic plants, subsequent damage and host responses to infestation. In *Thrips Biology and Management*. Edited by Parker BL, Skinner M, Lewis T. New York, Plenum; 1995:31-51.

30. Creelman RA, Mullet JE: Biosynthesis and action of jasmonates in plants. *Annu Rev Plant Physiol Plant Molec Biol* 1997, 48:355-381.

31. Delker C, Stenzel I, Hause B, Miersch O, Feussner I, Wasternack C: Jasmonate biosynthesis in *Arabidopsis thaliana* : enzymes, products, regulation. *Plant Biol (Stuttg.)* 2006, 8:297-306.

32. Thomma B, Eggermont K, Penninckx I, Mauch-Mani B, Vogelsang R, Cammue BPA, Broekaert WF: Separate jasmonate-dependent and salicylate-dependent defense-response pathways in Arabidopsis are essential for resistance to distinct microbial pathogens. *Proc Natl Acad Sci USA* 1998, 95:15107-15111.

33. Sasaki-Sekimoto Y, Taki N, Obayashi T, Aono M, Matsumoto F, Sakurai N, Suzuki H, Hirai MY, Noji M, Saito K, Takamiya K, Shibata D, Ohta H: Coordinated activation of metabolic pathways for antioxidants and defence compounds by jasmonates and their roles in stress tolerance in Arabidopsis. *Plant J* 2005, 44:653-668.

34. Schaller A, Stintzi A: Jasmonate biosynthesis and signaling for induced plant defense against herbivory. In *Induced Plant Resistance to Herbivory*. Edited by Schaller A. New York, Springer; 2008:349-366.

35. Browse J, Howe GA: New weapons and a rapid response against insect attack. *Plant Physiol* 2008, 146:832-838.

36. Chini A, Fonseca S, Fernández G, Adie B, Chico JM, Lorenzo O, Garcia-Casodo G, Lopes-Vidriero I, Lozano FM, Ponce MR, Micol JL, Solano R: The JAZ family of repressors is the missing link in jasmonate signaling. *Nature* 2007, 448:666-671.

37. Thines B, Katsir L, Melotto M, Niu Y, Mandaokar A, Liu G, Nomura K, He SH, Howe GA, Browse J: JAZ repressor proteins are targets of the SCFCOI1 complex during jasmonate signaling. *Nature* 2007, 448:661-665.

38. Yan Y, Stolz S, Chetelat A, Reymond P, Pagni M, Dubugnon L, Farmera EE: A downstream mediator in the growth repression limb of the jasmonate. *Plant Cell* 2007, 15:2866-2884.

39. Chung HS, Koo AJK, Gao X, Jayanty S, Thines B, Jones AD, Howe GA: Regulation and function of Arabidopsis *JASMONATE ZIM*-domain genes in response to wounding and herbivory. *Plant Physiol* 2008, 146:952-964.

40. Annadana S, Peters J, Gruden K, Schipper A, Outchkourov NS, Beekwilder MJ, Udayakumar M, Jongsma MA: Effects of cysteine protease inhibitors on oviposition rate of the western flower thrips, *Frankliniella occidentalis*. *J Insect Physiol* 2002, 48:701-706.

41. Botella MA, Xu Y, Prebha TN, Zhao Y, Narasimhan ML, Wilson KA, Nielsen SS, Bressan RA, Hasegawa PM: Differential expression of soybean cysteine proteinase inhibitor genes during development and in response to wounding and methyl jasmonate. *Plant Physiol* 1996, 112:1201-1210.

42. Belenghi B, Acconcia F, Trovato M, Perazzolli M, Bocedi A, Polticelli F, Ascenzi P, Delledonne M: AtCYS1, a cystatin from *Arabidopsis thaliana*, suppresses hypersensitive cell death. *Eur J Biochem* 2003, 270:2593-2604

43. Barth C, Jander G: Arabidopsis myrosinases TGG1 and TGG2 have redundant function in glucosinolate breakdown and insect defense. *Plant J* 2006, 46:549-562.

44. Shroff R, Vergara F, Muck A, Svatoš A, Gershenzon J: Nonuniform distribution of glucosinolates in *Arabidopsis thaliana* leaves has important consequences for plant defense. *Proc Natl Acad Sci* 2008, 105:6196-6201.
45. Konno K, Hirayama C, Nakamura M, Tateishi K, Tamura Y, Hattori M, Kohno K: Papain protects papaya trees from herbivorous insect: role of cysteine proteases in latex. *Plant J* 2004, 37:370-378.
46. Konno K, Ono H, Nakamura M, Tateishi K, Hirayama C, Tamura Y, Hattori M, Koyama A, Kohno K: Mulberry latex rich in antidiabetic sugar-mimic alkaloids forces dieting on caterpillars. *Proc Natl Acad Sci* 2006, 107:1337-1341.
47. Aharoni A, Giri AP, Deuerlein S, Griepink F, de Kogel WJ, Verstappen FW, Verhoeven HA, Jongsma MA, Schwab W, Bouwmeester HJ: Terpenoid metabolism in wild-type and transgenic Arabidopsis plants. *Plant Cell* 2003, 15:2866-2884.
48. Kappers IF, Aharoni A, van Herpen TWJM, Luckerhoff LLP, Dicke M, Bouwmeester HJ:Genetic engineering of terpenoid metabolism attracts bodyguards to Arabidopsis. *Science* 2001, 23:2070-2072.
49. Sánchez-Hernández C, López MG, Délano-Frier JP: Reduced levels of volatile emissions in jasmonate-deficient *spr2* tomato mutants favour oviposition by insect herbivores. *Plant Cell Environ* 2006, 29:547-557.
50. Schoonhoven LM, van Loon JJA, Dicke M: Host-plant selection: when to accept a plant.In *Insect-plant biology*. Oxford, Oxford university press; 2005:169-199.
51. Pauwels L, Morreel K, De Witte E, Lammertyn F, Van Montagu M, Boerjan W, Inzé D, Goossens A: Mapping methyl jasmonate-mediated transcriptional reprogramming of metabolism and cell cycle progression in cultured Arabidopsis cells. *Proc Natl Acad Sci* 2008, 105:1380 1385.
52. Narusaka M, Abe H, Kobayashi M, Kubo Y, Narusaka Y: Comparative analysis of expression profiles of counterpart gene sets between *Brassica rapa* and *Arabidopsis thaliana* during fungal pathogen *Colletotrichum higginsianum* infection. *Plant Biotechnol* 2006, 23:503-508.

53. Baldwin IT, Hamilton W: Jasmonate-Induced Responses of Nicotiana sylvestris Results in Fitness Costs Due to Impaired Competitive Ability for Nitrogen. *Journal of Chemical Ecology* 2000, 26:915-952.
54. Cipollini D: Consequences of the overproduction of methyl jasmonate on seed production, tolerance to defoliation and competitive effect and response of*Arabidopsis thaliana. New Phytologist* 2006, 173:146-153.
55. Seo HS, Song JT, Cheong J-J, Lee Y-H, Lee Y-W, Hwang I, Lee JS, Choi YD: Jasmonic acid carboxyl methyltransferase: A key enzyme for jasmonate-regulated plant responses. *Proc Natl Acad Sci* 2001, 95:9750-9754.
56. Thaler JS, Stout MJ, Karban R, Duffey SS: Jasmonate-mediated induced plant resistance affects a community of herbivores. *Ecological Entomology* 2001, 26:312-324.
57. Weigel D, Glazebrook J: How to grow Arabidopsis. In *Arabidopsis: A Laboratory Manual.* Edited by Weigel D, Glazebrook J. New York, Cold Spring Harbor Laboratory Press; 2002:1-18.
58. Murai T, Loomans AJM: Evaluation of an improved method for mass-rearing of thrips and a thrips parasitoid. *Entomol Exp Appl* 2001, 101:281-289.
59. Seo S, Okamoto M, Seto H, Ishizuka K, Sano H, Ohashi Y: Tobacco MAP kinase: a possible mediator in wound signal transduction pathways. *Science* 1995, 270:1988-1992.

Chapter 5

A Modular Cage System Design for Continuous Medium to Large Scale in Vivo Rearing of Predatory Mites (Acari: Phytoseiidae)

Juan Alfredo Morales-Ramos and Maria Guadalupe Rojas

USDA-ARS National Biological Control Laboratory, Biological Control of Pests Research Unit, 59 Lee Road, Stoneville, MS 38776, USA

ABSTRACT

A new stackable modular system was developed for continuous in vivo production of phytoseiid mites. The system consists of cage units that are filled with lima beans, Phaseolus lunatus, or red beans, P. vulgaris, leaves infested with high levels of the two-spotted spider mites, Tetranychus urticae. The cage units connect with each other through a connection cup, which also serves for monitoring and collection.

Predatory mites migrate upwards to new cage units as prey is depleted. The system was evaluated for production of Phytoseiulus persimilis. During a 6-month experimental period, 20, 894.9 ± 10,482.5 (mean ± standard deviation) predators were produced per week. The production consisted of 4.1 ± 4.6% nymphs and 95.9 ± 4.6% adults. A mean of 554.5 ± 59.8 predatory mites were collected per harvested cage and the mean interval length between harvests was 6.57 ± 6.76 days. The potential for commercial and experimental applications is discussed.

INTRODUCTION

Phytoseiid mites are very effective predators used mainly in biological control of spider mites, Tetranychus urticae (Koch); however, phytoseiids are known to provide effective control of other mite species and some insects like thrips and white flies [1]. Zhang [2] reported that at least 20 species of phytoseiids have been made commercially available and have been applied mainly on greenhouse plants. The phytoseiid that has been most widely mass-produced and sold commercially is Phytoseiulus persimilis Athias-Henriot. Phytoseiulus persimilisis an effective biological control agent of spider mites on vegetables in glasshouses [3–5] and growers around the world use P. persimilis to control T. urticae and other tetranychid mites on crops grown in greenhouses and in the field [6, 7]. Other phytoseiid species produced commercially and used in augmentative biological control of greenhouse pests include Neoseiulus cucumeris (Oudemans), N. barkeri Hughes, N. californicus (McGregor), N. fallacis (German), Iphiseius degenerans (Berlese), and Galendromus occidentalis (Nesbitt) [2].

Current methods of mass production of phytoseiid mites such as P. persimilis rely on greenhouse growth of bean plants for spider mite production and later inoculation with the predatory mite. A pure spider mite culture, free of predators, is also required for rearing. Infested leaves from the pure culture are used to infest bean plants in a different greenhouse. A series of greenhouse benches are inoculated at weekly intervals to provide continuous supply of prey. Predators are later introduced to bean plants heavily infested with spider mites and grown for 2-3 weeks. A section of the bench is harvested when it has reached the maximum predator density [8]. Introduction of P. persimilis into the infested beans requires perfect timing to allow maximum spider mite reproduction without losing the plants to the mite infestation [8]. Predator harvesting often exposes the predators to stressful conditions of starvation and many are lost to inefficient collection methods.

Enclosed rearing systems offer the potential of greater control of environmental conditions and better containment preventing excessive losses. Several methods for rearing phytoseiid mites in enclosed systems or cages by introducing prey have been proposed consisting of dishes with a central area limited by a channel filled with machine oil or other liquids [9–11].

Theaker and Tonks [12] reared P. persimilis in floating plastic leads positioned by magnets in the middle of a water-filled container to prevent mites from escaping. A similar method based on a plastic foam block or sponge positioned in the middle of a tray filled with water was described by Overmeer [13] to rear several species including P. persimilis, P. macropilis (Banks), Typhlodromus occidentalis Nesbitt, T. pyri Scheuten,Amblyseius (Neoseiulus) fallacis (German), A. potentillae (German), and N. cucumeris (Oudemans). A barrier is formed by placing wet tissue paper around the block with one side touching the water to maintain a continuous saturation [13]. McMurtry et al. [14] describe a method of mass rearing of phytoseiid mites by washing eggs and other spider mite stages from infested leaves. The washed spider mites are then fed to predatory mites reared using the paper-lined block in a water tray method as described by Overmeer [13]. Series of these trays are stacked inside shelved wood boxes. Shih [15] developed a method to separate the prey mites (T. urticae) from plant leaves and an apparatus which used pneumatic pressure to dispense a mix of the prey and corn pollen to rear Amblyseius womersleyi Schicha using the same lined semisubmerged block method.

Fournier et al. [16] proposed a cage system for rearing P. persimilis consisting of series of superimposing cylinders filled with bean leaves heavily infested by spider mites. New cylinders with infested leaves are added to the top of the series to supply new prey. The cylinders at the bottom are retired as predators move into cylinders with fresh prey [16]. Another cage system was described by Overmeer [13] consisting of two cardboard ice cream containers glued together and separated by a screen. Infested leaves are placed in the lower side and predators are introduced. New leaves are placed in the upper side where predators tend to move to find new prey. The whole system is flipped over to place new leaves while removing the old material [13].

While many enclosed rearing systems have been effective to mass-produce phytoseiid mites commercially, none of the existing enclosed systems can match the production capabilities of the open greenhouse rearing methods. An increasing level of sophistication will be required to reach a comparable level of production using enclosed systems. The objective of this study was to develop and test a refined enclosed rearing system based on the Fournier et al. [16] cage series.

MATERIALS AND METHODS

Rearing of the Prey

Spider mites, Tetranychus urticae Koch, were used as prey to feed the phytoseiid mites. Colonies of T. urticaewere established from commercial stocks provided by Syngenta Bioline, Oxnard, CA, and were reared on red kidney beans, Phaseolus vulgaris L., and lima beans, Phaseolus lunatus L., cultivars Fordhook 242 and Henderson in a greenhouse. The greenhouse was divided into two areas by using a clear polyethylene curtain. Lima beans were grown in one-half of the greenhouse using 60 × 20 × 20 cm polyethylene planters. The bottom of each planter was lined with 2.5 L perlite (Coarse, Sunshine, SunGro Horticulture, Bellevue, WA) to support and maintain humidity. A mixture of 2:1 potting soil (Moist control, Miracle-Gro Marysville, OH), vermiculite (Coarse, Sunshine, SunGro Horticulture, Bellevue, WA), and 5 g slow release fertilizer (N:P:K = 14:14:14) (vegetable and bedding, Osmocote, Marysville, OH) was mixed and then combined with an equal volume of a mixture of 20 g TeraGel (T-400, The Terawet Corporation, San Diego, CA), 0.5 g water soluble fertilizer (N:P:K = 24:8:16) (All-purpose plant food, Miracle-Gro, Marysville, OH), and 2.5 L tap water. The aqueous solution was allowed to equilibrate for 24 h until the water was fully absorbed by the TeraGel crystals and then it was homogeneously incorporated into the potting soil and vermiculite mixture using a gardening trowel. Seventy seeds were planted and kept for 10 days in each planter for germination. Ten days after germination, planters with young bean plants were transferred to the second half of the divided greenhouse and plants were then massively infested with T. urticae by placing leaves from heavily infested plants on top of them.

Spider mite infestation levels were allowed to increase for 5 days after their introduction. Fully infested plants were monitored daily to determine optimal infestation levels. Extreme infestation levels kill the bean plants inducing massive migration of spider mites. Infested bean leaves were collected when they were still alive (leaves still green) and sustain a high density of spider mites. Infested leaves were collected daily by cutting them manually using garden scissors and placing them in plastic boxes. Boxes with infested bean leaves were stored at 15°C for 1 to 7 days.

Predatory Mite Rearing and Cage Design

Although the rearing system presented herein is suitable for any phytoseiid predator of spider mites, P. persimilis was used as the basis to test the system. The rearing system is based on the same principles of Fournier et al.'s [16] stacked cage method, but we designed a unique modular structure of identical cage units. Stackable modular units were constructed from 473 mL Ziploc storage containers (Ziploc Twist'n Loc, S.C. Johnson & Son, Inc., Racine, WI), 250 mL plastic laboratory funnels (Fisherbrand 10-500-2, 10.5 cm dia. × 10.3 cm H), plastic 50 mL centrifuge tubes (Corning No. 430897), and 120 mL specimen containers (LSS number 9BC-135972) (Figure 1). Materials for cage construction were chosen based on the quality of a water-tight screwed cover closure. Snap closures tend to fail with continuous use and mites quickly find escape openings. The cage system consisted of 4 basic parts that were modified to fit together: a cage bottom (Figure 2(a)), a cage cover with funnel connection (Figure 2(b)), a connection cup (Figure 2(c)), and a multiuse funnel to serve as stand (Figure 2(d)). The covers of the Ziploc containers (Figure 1(c)) were cut to allow the insertion of the lab funnels to the cage covers (Figure 1(b)). The tips (narrow ends) of the funnels were cut to install "male" screw sections of centrifuge tubes to allow closure when required (Figures 2(b) and 2(d)). "Female" screw sections of the covers of specimen containers (Figure 1(d)) were cut and glued to the funnels and bottoms of the cage to allow connection with other cage units (Figures 2(a) and 2(b)). "Male" screw sections of centrifuge plastic tubes (Figure 1(a)) were also glued to the bottom of the connection cups (Figure 2(c)) to allow closure during mite collection and movement when connected to new cage units. Connection cups were also fitted with a second "male" screw section in the bottom allowing connection to both ends (Figure 3(c)). A cage unit consisted of bottom, top, and connection cup (Figure 3(a)). The system stand was used only in the starting cage (Figure 3(b)) and was fitted with a "male" instead of a "female" screw section from specimen containers (Figure 2(d)) to allow connection to the bottoms (Figure 2(a)) of the cages. Four circular windows (22 mm dia.) were cut on the sides of the cages bottom and four more (17 mm dia.) were cut on the sides of the funnels for ventilation. Only one circular window (10 mm dia.) was cut on one side of the connection cups to reduce excessive loss of moisture. Nylon screen 85 μm mesh (Small Parts Inc., U-CMN-85) was used to seal the windows preventing mites from escaping. Cage units were designed to fit together in a modular way by connecting the bottom to the connection cup (Figures 3(c) and 3(d)).

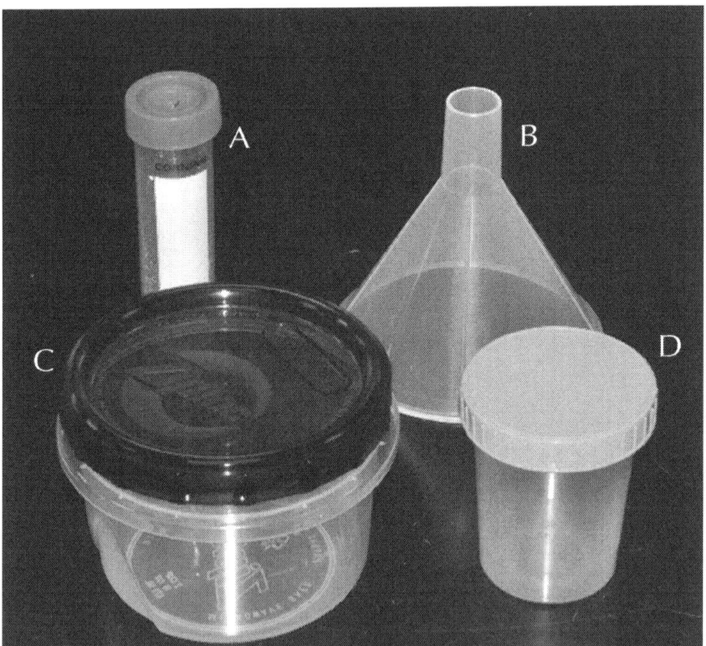

Figure 1: Materials used to construct the cage system. (A) Polypropylene 50 mL centrifuge tube with screwed cover, (B) polypropylene 250 mL lab funnel, (C) polypropylene Ziploc storage containers, and (D) high density polyethylene 120 mL specimen containers.

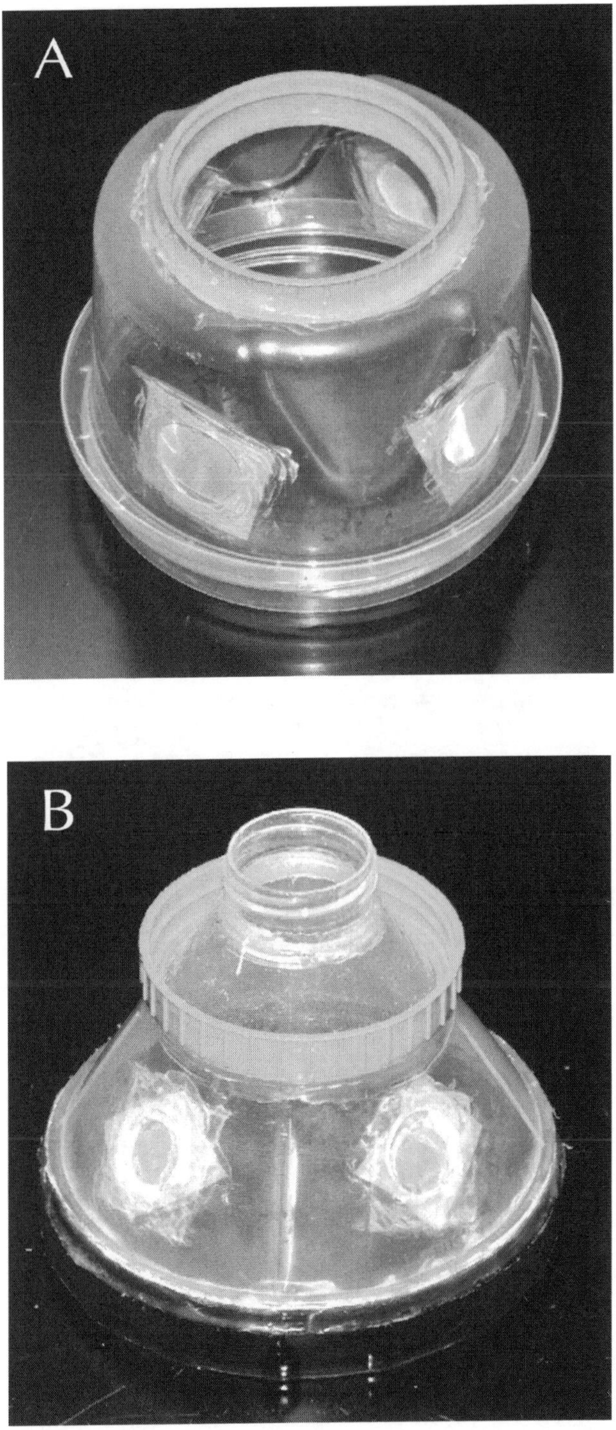
A
B

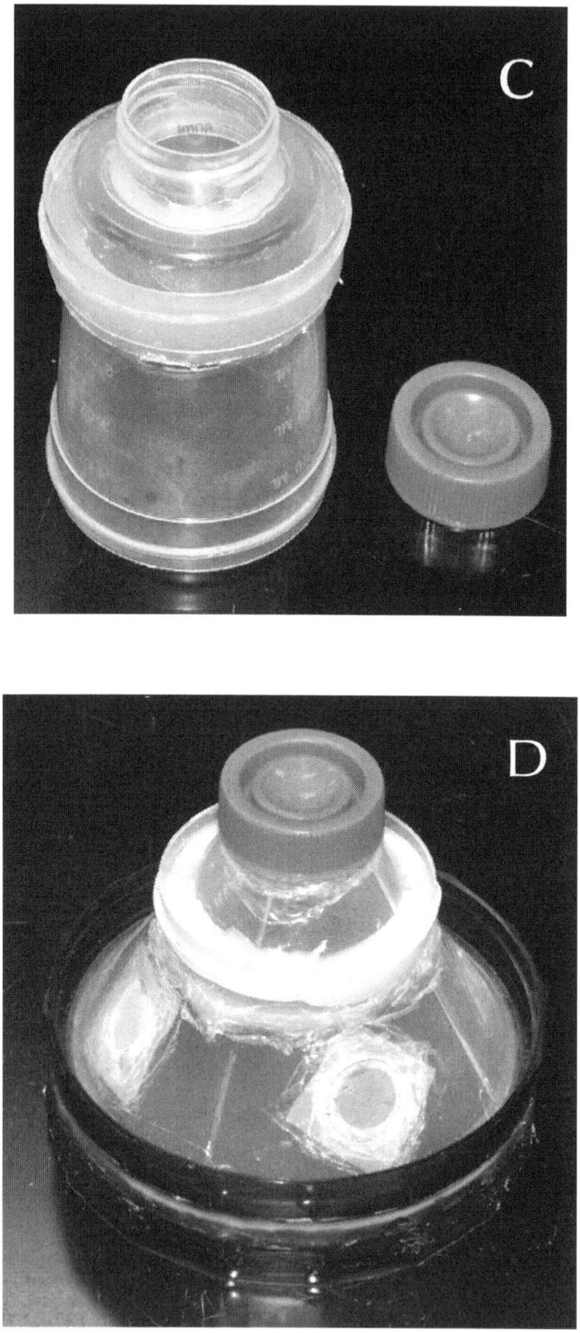

Figure 2: Basic cage system modular components. (a) Cage bottom, (b) cage cover, (c) connection cup, and (d) cage series stand.

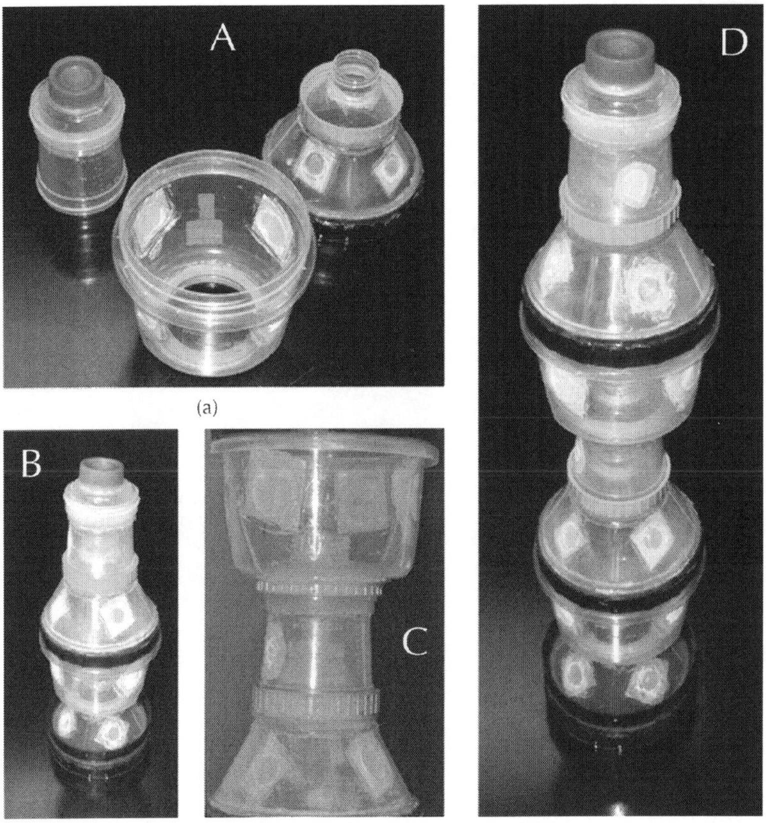

Figure 3: Cage system assembly. (a) Cage unit components, (b) cage unit assembled, (c) connection cup with second cage unit fitted, and (d) cage series assembly of two cage units.

Bean leaves heavily infested with T. urticae were placed inside each cage unit stacked vertically to allow mites to move up (Figures 4(a) and 4(c)). A cage series can be started by introducing a few adult predatory mites (10–100) into a cage unit newly filled with infested bean leaves. To start a cage series, a connection cup with mites is fitted to the bottom of a new cage unit (Figure 4(a) arrow and Figure 4(b)). This cup can later be replaced by the stand described in Figure 2(d) to provide better stability.

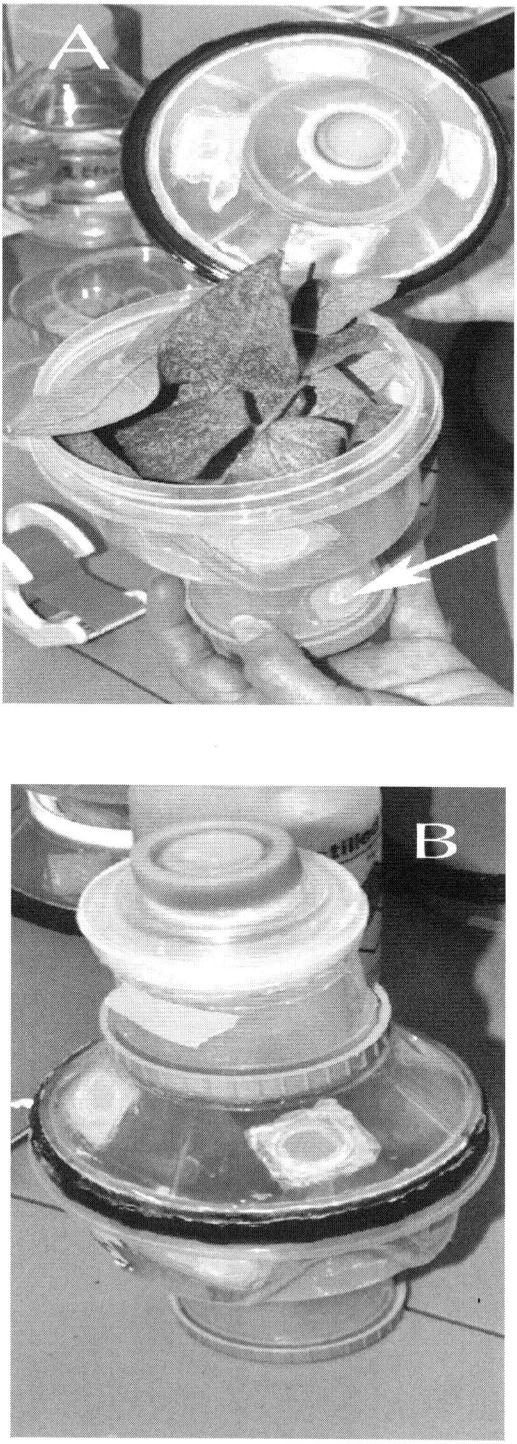
A
B

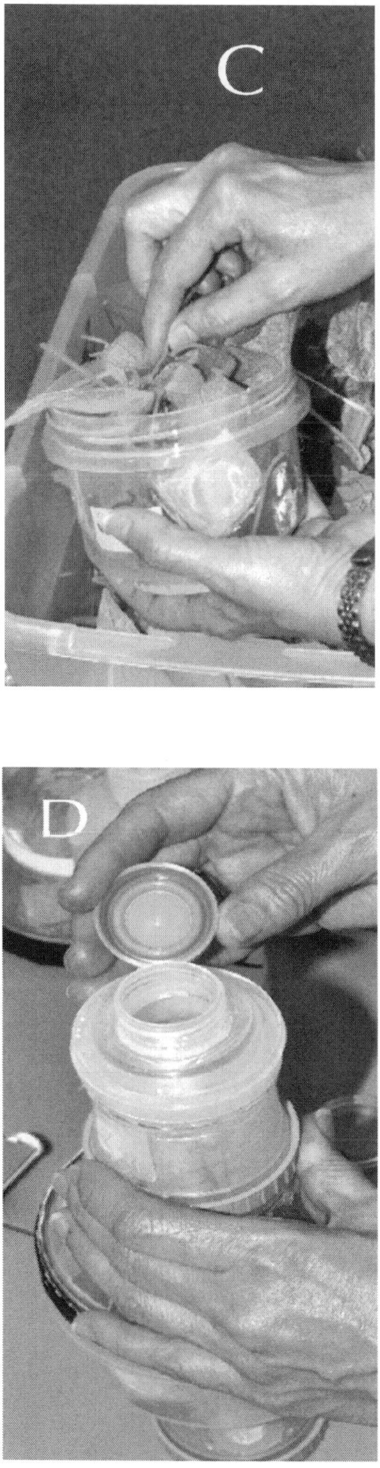
C
D

E

F

G

H

Figure 4: Cage system operation. (a) A cage series starts with a connection cup with predators connected to a cage bottom filed with spider mite-infested bean leaves. (b) The starting cage unit is closed and a connection cup is fitted. (c) When spider mites have been depleted by the predators, a new cage unit is prepared. (d) The cover of the connection cup is removed to allow predators to move to the new cage. (e) A new cage bottom is fitted to the connection cup. (f) The connection cup cover is placed inside the new cage. (g) The new cage is covered and (h) sealed, and (i) a new connection cup is fitted to the cover. (j) The process can be repeated by adding a third cage unit

when prey has been depleted. At this point, the bottom connection cup can be replaced by the stand piece.

After prey mites had been depleted, a new unit is attached to the top of the old unit by removing the cover of the connection cup (Figure 4(d)) to allow predators to move into the new unit. The cover (full of predators) is placed inside the new unit (Figures 4(e) and 4(f)) and the cage is closed with a new funnel (Figures 4(g) and 4(h)). A new connection cup is attached to the top of the funnel (Figure 4(i)) to complete the system (Figure 4(j)). Nymph and adult predators tend to migrate to the upper end of the cage series and accumulate in the uppermost connection cup feeding on migrating spider mites.

Evaluation

The cage system was evaluated using P. persimilis as model. Evaluation started on 1 January 2007 by establishing 21 cage series. Each cage series started with approximately 100 adult predators. New cage series were created using predatory mites produced by the initial series. Some cage series had to be terminated and replaced due to contamination by other predatory mite species (Neoseiulus sp.) from the greenhouse spider mite production. Series increased in number to a maximum of 48 at the end of the study on 1 July 2007. The study was conducted in an environmentally controlled room at 26 ± 1°C, 80 ± 5% RH, 14 h photophase, and 10 h scotophase.

Cage units were added to the top of each cage series as described above. Prey mites consisting of T. urticaewere reared as described above using P. lunatus Henderson variety. When the connection cup contained a visibly high density of predatory mites, the cup was quickly disconnected from the series, inverted, taped to make predator fall to the bottom, and filled with 70% ethanol to kill and preserve the predatory mites. Mites were counted and the numbers were recorded. Data consisting of days between harvest, collection date, cage series, and number of P. persimilis collected were recorded. Data were analyzed using single-variable statistics to determine means of mite production per week, mites produced per cage, and mean time from initiation to collection.

In this study, predatory mites were collected and killed with 70% ethanol in order to obtain precise numbers. However, live predators can be quantified while alive using less precise methods. One method is based on weight: first, a mean of individual weight is determined by weighing groups of mites in a precision balance; second, an empty collection cup (with cover) is weighed and used to collect predatory mites by attaching it to the top of the cage system. The cup full of mites can be closed (with the previously weighed cover) and weighed a second time. The weight of the live mites can be determined by subtracting the weight of the empty cup from the weight of the full cup. Another method consists of determining the number of mites fitting in a given volume. Mites can be forced by gentle vacuum into receptacles with known volume. When filled, the receptacle can be emptied by reversing the airflow.

RESULTS AND DISCUSSION

During the six-month evaluation period, the mean weekly production was 20,894.9 ± 10 ,482.5 (mean ± standard deviation) P. persimilis. Production consisted of 4.1 ± 4.6 % deutonymphs and 95.9 ± 4.6 % adults. Overall production mean was 554.5 ± 59.8 mites per harvested cage and a mean of 36.7 ± 17.0 cages were harvested per week. The mean interval of time between harvests was 6.57 ± 6.76 days; however, the length of the harvest intervals did not have a normal distribution and the median was 4 days and the 75% quartile was 5 days (Figure 6). The top quartile consisted of intervals ranging from 6 to 62 days in length and the 90% percentile was 14 days. Based on this analysis, the time interval between harvests should not exceed 14 days. If by 14 days the population of predators is still too low to justify harvest, the cage series should be terminated and a new series should be generated from predator production of a healthy series.

The connection cup usually contains large quantities of spider mites, which constantly migrate when they are in high densities. As predatory mites increase in numbers, they consume all the prey in the connection cup. Juvenile predatory mites (young adults and deutonymphs) tend to rapidly migrate upwards to the new cage unit. Gravid adult females usually remain in the bottom cage unit until they oviposit. As the last eggs are oviposited and prey is depleted, females move upwards to the new cage unit. As the population of predatory mites increases, it becomes necessary to add new cage units within increasingly shorter periods of time. Empty cage units at the bottom can be removed after all the eggs have hatched and the juveniles have moved to new cage units. The connection cup at the top of the series serves as an indicator of predatory mite population. The predators in the connection cup may be recycled by fitting a new cage unit or harvested by removing the connection cup and closing it with an unmodified cover of a specimen container. Decision to harvest the predatory mites depends on the density of juveniles in the connection cup (Figure 5). Once the predatory mite population is well established in a cage series, it becomes necessary to harvest the predatory mites every 2 to 4 days depending on the quality of prey provided. Harvested mites can be used to start new cage series, for field releases, or for use in experiments. A cage series can be continuously producing predatory mites indefinitely as long as new cage units are added to the top of the series.

Figure 5: Connection cup with P. persimilis after prey had been depleted. A piece of paper can be introduced to the connection cup to increase the surface area.

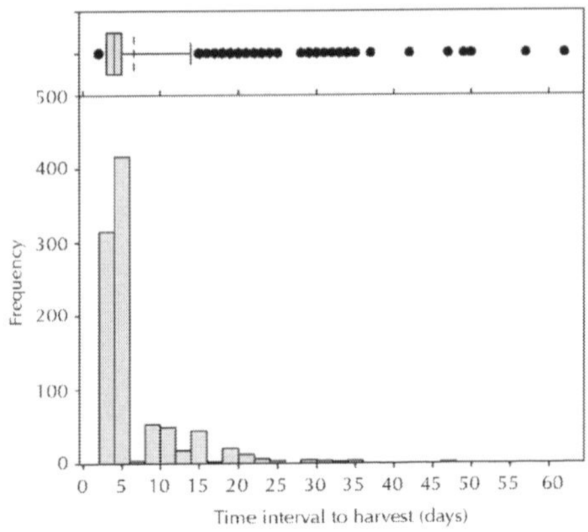

Figure 6: Frequency distribution of predator harvest intervals. Box plot (top): bars represent 25% and 75% quartiles, line between bars represents the median, dashed line represents the mean, bracket represents the 90% percentile, and dots represent outliers. Bar plot (bottom): bars represent 2-unit classes.

Peak weekly production occurred during the week between March 11 and 18 with a total of 38,097 adults and nymphs harvested (Figure 7(a)). During this period, 41 cage series were in operation (Figure 7(b)). Mean production per harvested cage was more or less consistent during the experimental period (Figure 7(c)). The total weekly production dropped sharply by the end of the experiment even as the number of series increased; however, the number of mites per harvested cage was increasingly consistent evidenced by the decrease of the standard deviation (Figure 7(c)).

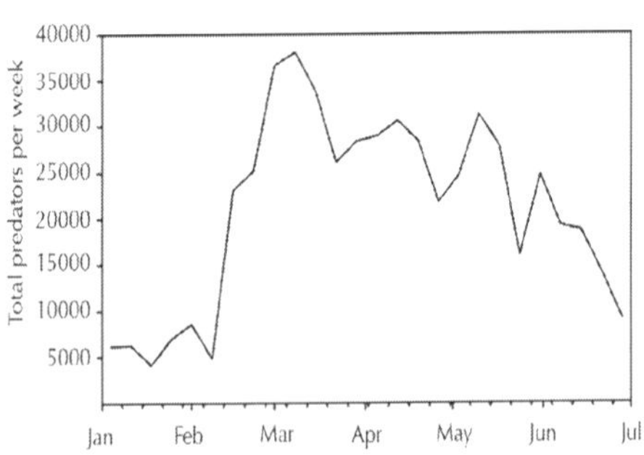

(a)

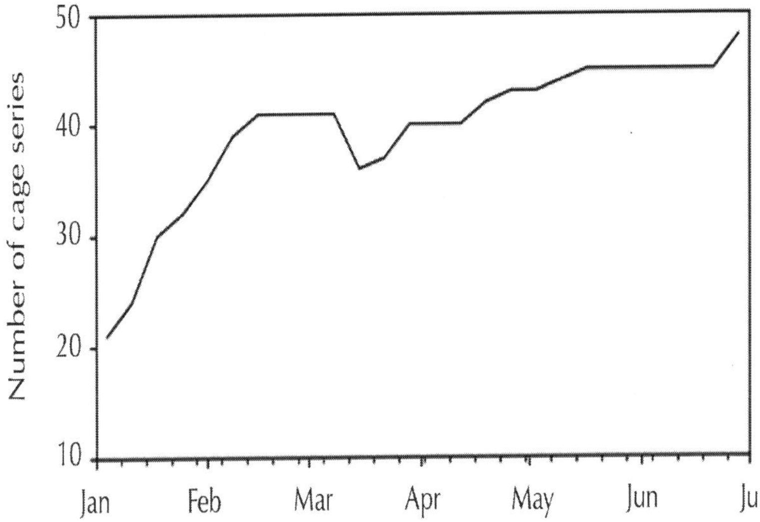
50
40
30
20
10
Number of cage series
Jan
Feb
Mar
Apr
May
Jun
Jul

(b)

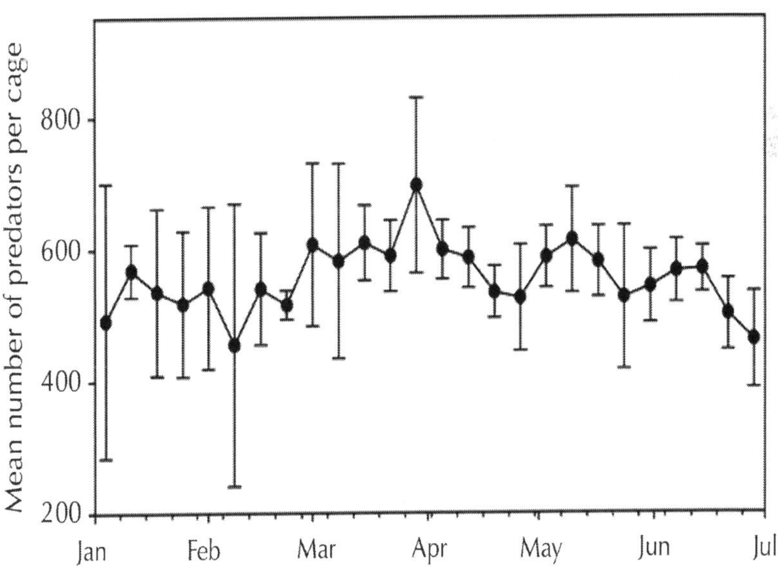
800
600
400
200
Mean number of predators per cage
Jan
Feb
Mar
Apr
May
Jun
Jul

(c)

Figure 7: Production of P. persimilis. (a) Total weekly production. (b) Number of cage units in production. (c) Means of predators produced per harvested cage; brackets represent standard deviation.

A study of three-trophic-level impact of secondary chemicals present in lima beans showed that high levels of linamarin present in Henderson lima beans tend to accumulate in the predatory mites after several generations [17]. Accumulation of linamarin may have been the reason the production dropped by the end of the study even as the number of cage series increased. The use of this variety is not recommended for continuous production of phytoseiid mites. Fordhook 242 lima beans provide a better alternative because they contain only trace amounts of linamarin (M. G. Rojas unpublished). Henderson lima beans were selected for this experiment because they are easy to grow and the size of their leaves is optimal to fit inside the cages described in this study. Fordhook 242 lima beans have larger leaves, which must be folded or cut to fit in the cages. Another good choice of host plant is red beans, which have similar leaf sizes to those in Henderson lima beans. However, a system could be constructed with larger cage units providing more space to accommodate larger leaves.

The size limits for the system have not been determined, but 8 liter (2-gallon) size units have been constructed and tested successfully (Figure 8). In theory, the system can be scaled up to accommodate production levels of millions of mites per week. Potential size limits include the structural integrity of currently available materials taking into account the weight of the leaves that must be held by the cage units. The tightness of the closures between cage unit connections can be more difficult as cage size increases. The tolerance of closures between cage units cannot exceed 100 μm to contain the mites within the cages, and this becomes increasingly difficult as the size of the screw cups increases.

Figure 8: Larger size cage system design.

CONCLUSIONS

The modular cage system presented in this study has been shown to be a consistent and robust method to produce phytoseiid mites. The system is particularly suitable for medium to large scale rearing of P. persimilis. This system provides a good alternative for phytoseiid mites rearing and potentially can be scaled up for mass production.

ACKNOWLEDGMENTS

The authors thank D. Cahn, Syngenta Bioline, for providing the rearing stocks of P. persimilis and T. urticaeand partial funding through CRADA agreement 58-3 K95-0-1428. They acknowledge the peer reviewers for their comments on an earlier version of this paper. The United States Government has the right to retain a nonexclusive, royalty-free license in and to any copyright of this paper. The mention of a commercial or proprietary product does not constitute an endorsement of the product by the United States Department of Agriculture (USDA). USDA is an equal opportunity provider and employer.

REFERENCES

1. K. J. F. Bolckmans, "Mass-rearing phytoseiid predatory mites," in Proceedings of the Working Group AMRQC, C. van Lenteren, P. DeClercq, and M. W. Johnson, Eds., vol. 3, pp. 12–15, Bulletin IOBC Global, 2007.
2. Z. Zhang -Q, Mites of Greenhouses Identification, Biology and Control, CABI, Oxon, UK, 2003.
3. D. A. Chant, "An experiment in biological control of Tetranychus telarius (Linneaus) (Acarina: Tetranychidae) in a greenhouse using the predacious mite Phytoseiulus persimilis Athias-Henriot (Phytoseiidae)," The Canadian Entomologist, vol. 93, no. 6, pp. 437–443, 1961.
4. T. J. Legowski, "Experiments on predator control of the glasshouse red spider mite on cucumbers," Plant Pathology, vol. 15, pp. 34–41, 1966.
5. N. French, W. J. Parr, H. J. Gould, J. J. Williams, and S. P. Simmonds, "Development of biological methods for the control of Tetranychus urticae on tomatoes using Phytoseiulus persimilis," Annals of Applied Biology, vol. 83, no. 2, pp. 177–189, 1976.
6. J. C. van Lenteren, "Commercial availability of biological control agents," in Quality Control and Production of Biological Control Agents: Theory and Testing Procedures, J. C. van Lenteren, Ed., pp. 167–179, CABI, Oxon, UK, 2003.
7. J. C. van Lenteren, "The state of commercial augmentative biological control: plenty of natural enemies, but a frustrating lack of uptake," BioControl, vol. 57, no. 1, pp. 1–20, 2012.
8. L. A. Gilkeson, "Mass rearing of phytoseiid mites for testing and commercial application," in Advances in Insect Rearing for Research and Pest Management, T. E. Anderson and N. C. Leppla, Eds., pp. 489–506, Westview Press, Boulder, Colo, USA, 1992.
9. S. S. Kamburov, "Methods of rearing and transporting predacious mites," Journal of Economic Entomology, vol. 59, pp. 875–877, 1966.

10. G. T. Scriven and J. A. McMurtry, "Quantitative production and processing of tetranychid mites for large-scale testing of predator production," in Journal of Economic Entomology, vol. 64, pp. 1255–1257, 1971.
11. T. Kostiainen and M. A. Hoy, "Egg-harvesting allows large scale rearing of Amblyseiulus finlandicus(Acari: Phytoseiidae) in the laboratory," Experimental & Applied Acarology, vol. 18, pp. 155–165, 1994.
12. T. L. Theaker and N. V. Tonks, "A method of rearing the predaceous mite Phytoseiulus persimilis(Acarina: Phytoseiidae)," Journal of the Entomological Society of British Columbia, vol. 74, pp. 8–9, 1977.
13. W. P. J. Overmeer, "Rearing and handling," in Spider Mites: Their Biology, Natural Enemies, and Control, W. Helle and M. W. Sabelis, Eds., pp. 161–170, Elsevier, Amsterdam, The Netherlands, 1985.
14. J. A. McMurtry, G. T. Scriven, S. N. Newberger, and H. G. Johnson, "Methodologies of rearing, introducing, and establishing phytoseiid mites," in Proceedings of the ADAP Crop Protection Conference, vol. 134 of Research Extension, pp. 104–110, HITAHR, Honolulu, Hawaii, USA, 1989.
15. C. I. T. Shih, "Automatic mass-rearing of Amblyseius womersleyi (Acari: Phytoseiidae)," Experimental and Applied Acarology, vol. 25, no. 5, pp. 425–440, 2001.
16. D. Fournier, P. Millot, and M. Pralavorio, "Rearing and mass production of the predatory mitePhytoseiulus persimilis," Entomologia Experimentalis et Applicata, vol. 38, no. 1, pp. 97–100, 1985.
17. M. G. Rojas and J. A. Morales-Ramos, "Tri-trophic level impact of host plant linamarin and lotaustralin on Tetranychus urticae and its predator Phytoseiulus persimilis," Journal of Chemical Ecology, vol. 36, no. 12, pp. 1354–1362, 2010.

Chapter 6

Management of Mango Hopper, Idioscopus clypealis, Using Chemical Insecticides and Neem Oil

S. M. Adnan[1], M. M. Uddin[1], M. J. Alam2, M. S. Islam[2], M. A. Kashem[2], M. Y. Rafii[3, 4], and M. A. Latif[4, 5]

[1]Bangladesh Agricultural University, Mymensingh 2202, Bangladesh

[2]Bangladesh Institute of Nuclear Agriculture, Mymensingh 2202, Bangladesh

[3]Institute of Tropical Agriculture (ITA), University Putra Malaysia, 43400 Serdang, Selangor, Malaysia

[4]Department of Crop Science, Faculty of Agriculture, University Putra Malaysia, 43400 Serdang, Selangor, Malaysia

[5]Bangladesh Rice Research Institute, Gazipur 1701, Bangladesh

ABSTRACT

An experiment was conducted in Field Laboratory, Department of Entomology at Bangladesh Agricultural University, Mymensingh, during 2013 to manage the mango hopper, Idioscopus clypealis L, using three chemical insecticides, Imidacloprid (0.3%), Endosulfan (0.5%), and Cypermethrin (0.4%), and natural Neem oil (3%) with three replications of each. All the treatments were significantly effective in managing mango hopper in comparison to the control. Imidacloprid showed the highest efficacy in percentage of reduction of hopper population (92.50 ± 9.02) at 72 hours after treatment in case of 2nd spray. It also showed the highest overall percentage of reduction (88.59 ± 8.64) of hopper population and less toxicity to natural enemies including green ant, spider, and lacewing of mango hopper. In case of biopesticide, azadirachtin based Neem oil was found effective against mango hopper as 48.35, 60.15, and 56.54% reduction after 24, 72, and 168 hours of spraying, respectively, which was comparable with Cypermethrin as there was no statistically significant difference after 168 hours of spray. Natural enemies were also higher after 1st and 2nd spray in case of Neem oil.

INTRODUCTION

Mango (Mangifera indica Linn.) is a very important and popular fruit in the world. It is the choicest fruit of the subcontinent and is known as king of all fruits. Its popularity is mainly due to its excellent flavour, delicious taste, and high nutritive value being rich in vitamins A and C. Its origin is believed to be south Asia where it has been cultivated for the last four thousand years [1]. Now it is a commercially cultivated important fruit of this subtropical region particularly Bangladesh, India, and Pakistan. But production of mango is enormously handicapped by the ravages of insect pests from seedling to their maturity. More than 300 insect pests have been recorded to attack mango crop in different regions of world [2]. Among the mango pests, Mango hopperIdioscopus clypealis (Lethierry) is one of the most serious and widespread pests throughout the country, which causes heavy damage to mango crop. Both the nymphs and adults of the hoppers puncture and suck the sap from tender shoots, inflorescences, and leaves of mango crop, which cause nonsetting of flowers and dropping of immature fruits, thereby reducing the yield. Hoppers also excrete a secretion, called honey dew. In moist weather, it encourages the development of fungi like Meliola mangiferae (Earle), resulting in growth of sooty mould on dorsal surface of leaves, branches, and fruits. This black coating interferes with the normal photosynthetic activity of the plant, ultimately resulting in nonsetting of flowers and dropping of immature fruits. This damage is called honey dew disease. On heavily infested trees, crop losses of 50% or more have been recorded [2]. In the past its control was based purely on chemicals especially synthetic insecticides. But nonjudicious application of highly toxic and persistent insecticides is causing several problems such as disrupting natural enemy complexes, development of insecticide resistance, secondary pest outbreak, pest resurgence, and environmental pollution [3]. To solve these problems farmers can shift from the unilateral reliance on insecticide use to alternative approaches. In this situation, biodegradable substitutes are now being strongly conceived by all scientists of the world. Biologically active natural plant products may play a significant role in this regard as they are environmentally safe, biodegradable, and cost effective. A large number of investigators isolated and identified several chemical compounds from leaves and seeds of many plant species and screened out many insect feeding deterrents and growth inhibitors [4]. Among them Neem based products have extensively been used and have proved their pest control efficacy against several insect pests both in field and storage. However, exploration on the use of botanicals against mango pests is scanty in Bangladesh. Under this circumstance, the present research was undertaken to manage mango hopper, I. clypealis (Lethierry), using three nonpersistence chemical insecticides, namely, Imidacloprid, Endosulfan, Cypermethrin, and natural product Neem oil as well as to assess their toxic effects on natural enemies of mango hopper.

MATERIALS AND METHODS

The experiment was conducted in the Entomology Field Laboratory at Bangladesh Agricultural University, Mymensingh. The experiment was performed following randomized complete block design (RCBD) having five treatments including control with three replications of each treatment. Five inflorescences from five different branches of the same tree were selected alternatively from top, middle, and bottom and were sprayed with each treatment. Selection of inflorescence was done as modified method developed by [5]. The treatments were three insecticides, namely, Imidacloprid, Endosulfan, Cypermethrin, and natural Neem oil. A control treatment was always maintained with three replications. Insecticides were sprayed as recommended dose and Neem oil was used as 3% concentration (Table 1). Application of treatments and collection of data were done before 10 a.m. Data were counted on the number of hopper per inflorescence. Pretreatment data were taken just before and posttreatment data were recorded after 24, 72, and 168 hours of spraying. The percentage reduction of insect population was computed using Henderson-Tilton's formula, that is, %efficacy = $[1-Ta/Ca \times Cb/Tb] \times 100$ [6], where, Tb is the infestation in the treated plot before treatment, Ta is the infestation in the treated plot after treatment, Cb is the infestation in the control plot before treatment, and Ca is the infestation in the control plot after treatment.

Table 1: Insecticides used in efficacy trials against mango hopper

Pesticides		Type	Source	Dose
Common name	Trade name			
Imidacloprid	Imidacloprid 25 WP	Neonicotinoid	Arysta LifeScience	0.3%
Endosulfan	Thiodan 35 EC	Organochlorine	Bayer CropScience	0.5%
Cypermethrin	Cypermethrin 20 EC	Pyrethroid	Syngenta	0.4%
Neem oil		Botanical	Laboratory of Bangladesh Agricultural University (BAU)	3%

Data were analyzed following ANOVA using statistical package SPSS (version 14.0). Significant differences among the means of different treatments were tested using Duncan multiple range test (DMRT). Efficacy of insecticides and Neem oil were tested on the basis of percentage reduction of hopper population. Along with this, effects of selected insecticides and Neem oil on natural enemies were assessed by taking pretreatment data just before spray and posttreatment data at 24 HAT of 1st, 2nd, and 3rd spray.

RESULTS AND DISCUSSION

Effect of Three Synthetic Insecticides in Controlling Mango Hopper

Reduction in population indicated that all the tested insecticides were effective against mango hopper (Table2). The percentage reduction over pretreatment in comparison with control plot was higher with Imidacloprid and Endosulfan followed by Cypermethrin and Neem oil. There was significant difference () among the treatments as overall efficacy. In case of conventional insecticides, Imidacloprid was found highly effective with 83.63% reduction after 24 hours of first spray with increasing trend in efficacy as 89.97% after 72 hours; however its effectiveness decreased at 168 hours to 85%, whereas Endosulfan proved to be a good controlling agent with 75.79% reduction at 24 hours with decreasing trend after 72 and 168 hours of spray with 72.48 and 67.47% reduction, respectively. Cypermethrin was comparatively less effective than Imidacloprid and Endosulfan as it gave 64.40% reduction after 24 hours though it increased after 72 hours to 68.23% reduction but again after 168 hours of first spray 57.95% reduction was observed. Similar results were reported by [7] with significant dominance of Imidacloprid among different pesticides against okra jassid (Amrasca biguttula biguttula) [8]. The performance of different doses of Imidacloprid as an agent of seed and root treatment was assessed and found effective up to 45 days after treatment against chilli thrips. It was found that Imidacloprid is the most effective against mustard aphid compared to Endosulfan and Neem oil [9], whereas [10] found Endosulfan very effective against jassids on okra crop. The effectiveness of Endosulfan and Cypermethrin against aphid on okra and brinjal crops at different time intervals was evaluated and it was found that Endosulfan is more effective than Cypermethrin on brinjal crop and vice versa in case of okra crop [11, 12].

Table 2: Percentage of efficacy of three insecticides against mango hopper

Treatment	Percentage of population reduction			
	24h	72h	168h	Mean
1st spray				
Imidacloprid	83.63 ± 8.91a	89.97 ± 10.06a	85.05 ± 10.94a	86.22 ± 9.13a
Endosulfan	75.79 ± 9.06ab	72.48 ± 9.42ab	67.47 ± 11.71ab	71.91 ± 9.49b
Cypermethrin	64.40 ± 10.98b	68.23 ± 11.72b	57.95 ± 10.20b	63.53 ± 10.53bc
Neem oil	48.35 ± 7.91c	60.15 ± 10.84b	56.54 ± 10.65b	55.01 ± 9.89c
2nd spray				
Imidacloprid	88.79 ± 1015a	92.50 ± 9.02a	90.12 ± 9.85a	90.47 ± 8.54a
Endosulfan	80.61 ± 10.48ab	72.48 ± 9.42ab	69.25 ± 9.11b	75.04 ± 9.84b
Cypermethrin	66.77 ± 9.26bc	70.85 ± 9.81bc	66.30 ± 9.13b	67.97 ± 8.43b
Neem oil	49.54 ± 9.42c	58.88 ± 11.81c	57.07 ± 11.93b	55.16 ± 10.54c
3rd spray				
Imidacloprid	88.42 ± 10.04a	90.18 ± 9.87a	88.65 ± 1.06a	89.08 ± 8.69a
Endosulfan	75.62 ± 11.17ab	72.97 ± 10.08ab	70.44 ± 11.84b	73.01 ± 9.83b
Cypermethrin	62.59 ± 11.89bc	66.45 ± 9.19b	60.25 ± 9.20b	63.10 ± 9.22bc
Neem oil	52.31 ± 10.30c	62.12 ± 11.81b	58.23 ± 12.90b	57.55 ± 10.41c
Overall percentage of efficacy				
Imidacloprid	88.59 ± 8.64a			

Endosulfan	73.32 ± 9.43b			
Cypermethrin	64.87 ± 9.33c			
Neem oil	55.91 ± 9.95d			

Values sharing the same letter(s) in a column are not significantly different at p=0.05.

The effect of Endosulfan and azadirachtin was studied and azadirachtin was found to be moderately effective against brinjal shoot and fruit borer when used alone, whereas it varied in efficacy when used in combination such as Endosulfan + Bt (Bacillus thuringiensis) and azadirachtin + Bt [13]. However, in any case azadirachtin was found comparatively less effective than Endosulfan. Whereas similar combinations of pesticides along with Cypermethrin against aphids and jassids of okra and observed significant dominance of some pesticides among others, Cypermethrin showed moderate but yet more effective than Neem product (azadirachtin).

Effect of Natural Neem Oil in Controlling Mango Hopper

Unlike synthetic pesticides plant based pesticides have diverse pest control properties. Plant products affect different physiological processes in insects like metamorphosis including insect growth regulation, adult fertility, and toxicity and also have antifeedant and oviposition deterrent effects [14].

It was reported that they are environmental friendly; therefore, they seem to have some superiority over synthetic pesticides [15]. Moreover, a variety of plant species are available with diverse types of controlling effects as over 2400 plants have been identified with pest control properties in this respect [16].

In case of biopesticides, azadirachtin based Neem oil was found effective against Mango hopper at the rate of 48.35, 60.15, and 56.54% reduction after 24, 72, and 168 hours, respectively, which was comparable with Cypermethrin as there was no statistically significant difference after 168 hours of spray () (Table 2). The efficacy of different Neem based readymade products was evaluated and observed to be comparatively less effective against sucking pests [17] than Endosulfan but proved superior to untreated plot. The work also supports our findings [18] and the authors tested the field efficacy of azadirachtin-A, its stable derivative tetrahydroazadirachtin-A (THA), and NeemAzal (NZ) pesticides, in comparison with Endosulfan against the complex pests of okra including jassid and whitefly, and found azadirachtin-A effective for up to 7 days whereas THA had potentiality to control the pests for up to 10 days.

Endosulfan was found to be most effective with 82.9% reduction in whitefly population followed by THA, Aza-A, and NZ (60%, 58.7% and 57.5%). Against jassid, it was reduced by 62% with Endosulfan followed by THA, Aza-A, and NZ as 40.2, 35.1, and 31%, respectively. The effectiveness of Biosal (Neem formulation) in comparison with Endosulfan and Profenofos against jassid on brinjal at different time intervals was evaluated and a moderate effect of Biosal against jassid was found with 47% mortality [19]. Crude extracts of Neem and Margosan-O (pesticide of Neem origin from USA, containing 0.3% azadirachtin as active ingredient) were tested in comparison with Malathion 57 EC [20] against white fly on brinjal and effectiveness was found as Malathion > Margosan > crude Neem extracts after 48 hours, while after 96 hours crude Neem extracts were more persistent than Margosan and Malathion, respectively. Similarly different attempts were made to test the efficacy of Neem preparations against aphid and whitefly [21, 22]. Neem extract was found effective but inferior to Imidacloprid against the spread of okra yellow vein mosaic virus by controlling whitefly population [23], whereas Neem extract was found more effective against jassid, white fly, and thrips on cotton as compared to Perfekthion which lost its efficacy after 4 days, while Neem product was persistent for up to 6 days and was much safer and nonpolluting. In our present study, Neem oil was also effective for up to 7 days [23]. Reference [24] reported that significant reduction in the population of jassid, whitefly, and thrips on cotton was up to 168 hours when Neem oil was used as 2% and Neem seed water extract as 3% but efficacy declined at 336 hours. Neem oil was effective with 56, 54, and 57% reduction in the population of jassid, whitefly, and thrips, respectively, while Neem seed water extract was relatively less effective with 49, 46, and 54% reduction against three insects. Imidacloprid, Endosulfan, Cypermethrin, and Neem oil showed the average reduction (overall performance) of hopper population in three sprays as 88.59, 73.32, 64.87, and 55.91%, respectively.

There was a significant variation among the treatments considering the number of sprays but they were statistically identical in respect of time after spray (Table 3).

Table 3: Efficacy of three insecticides and Neem oil: effect of spray and time

Treatment	Percentage of population reduction			
	Imidacloprid	Endosulfan	Cypermethrin	Neem oil
Spray				
Spray 1	86.22 ± 9.13a	71.91 ± 9.49b	63.53 ± 10.53bc	55.01 ± 9.89c
Spray 2	90.47 ± 8.54a	75.04 ± 9.84b	67.97 ± 8.43b	55.16 ± 10.54c
Spray 3	89.08 ± 8.69a	73.01 ± 9.83b	63.10 ± 9.22bc	57.55 ± 10.41c
Time (h)				
24h	86.95 ± 8.77a	77.34 ± 9.23a	64.59 ± 9.50a	50.07 ± 8.12a
72h	90.88 ± 8.45a	73.57 ± 8.58a	68.57 ± 9.12a	60.38 ± 9.39a
168h	87.94 ± 9.20a	69.06 ± 9.58a	61.50 ± 9.05a	57.28 ± 10.15a

Values sharing the same letter(s) in a column are not significantly different at p=0.05.

Effect of Three Chemical Insecticides and Neem Oil on Natural Enemies

Effect of different insecticides and Neem oil on natural enemies of mango hopper, namely, green ant, spider, and lacewing varied significantly (Table 4). Among the four treatments Neem oil showed least toxicity on all the natural enemies (green ant, spider, and lacewing). Imidacloprid and Cypermethrin were moderately toxic to the natural enemies of mango hopper, whereas Endosulfan reduced hopper population significantly but it was highly toxic to the natural enemies of mango hopper (Table 4). The highest number of natural enemies' population was always observed in control plot.

Table 4: Effect of three chemical insecticides and Neem oil on natural enemies

Treatments	Number of green ants				Number of spiders				Number of lacewings			
	Before spray	At 1st spray	At 2nd spray	At 3rd spray	Before spray	At 1st spray	At 2nd spray	At 3rd spray	Before spray	At 1st spray	At 2nd spray	At 3rd spray
Imidacloprid	4.33	2.95bc	3.67b	3.82b	3.23	2.87b	3.45b	2.31b	0.87	0.74b	0.56b	0.83b
Endosulfan	3.98	1.14c	1.01c	1.21c	4.78	0.92c	0.87c	0.68c	0.65	0.13c	0.06c	0.19c
Cypermethrin	4.12	2.65bc	3.34b	5.54ab	5.98	2.12b	1.98b	2.43b	1.87	1.62b	1.74b	1.41b
Neem oil	4.89	3.18b	4.11b	3.65b	3.75	4.34b	2.76b	2.13b	1.43	0.98b	1.30b	0.98b
Control	4.98	7.65a	6.98a	7.76a	3.62	5.43a	7.98a	4.61a	1.32	2.97a	2.65a	2.13a
Level of significance	**				**				**			

Values sharing the same letter(s) in a column are not significantly different at P=0.05.

This finding was in agreement with [25] who reported that Endosulfan was highly toxic to the predators of potato leaf hopper. It could be concluded from the findings of the present study that sole dependency on conventional insecticides may easily be modified by incorporating Neem oil in an environment friendly management program for mango hopper.

CONCLUSIONS

Chemical insecticides, Imidacloprid (0.3%), Endosulfan (0.5%), Cypermethrin (0.4%), and Neem oil (3%) were effective in managing mango hopper in comparison to the control. Imidacloprid showed the highest efficacy in percentage of reduction of hopper population at 72 hours after treatment in case of 2nd spray. It also showed the highest overall percentage of reduction of hopper population and less toxicity to natural enemies of mango hopper. In case of biopesticide, azadirachtin based Neem oil was found effective against Mango hopper as 48.35, 60.15, and 56.54% reduction after 24, 72, and 168 hours, respectively. The population of natural enemies was also found higher in case of Neem oil even after 1st and 2nd spray. So, it could be concluded that sole dependency on conventional insecticides may easily be modified by incorporating Neem oil as an environment friendly management program for mango hopper.

REFERENCES

1. D. K. Salunkhe and B. B. Desai, Post-Harvest Technology of Fruits, vol. 1, CRC Press, Boca Raton, Fla, USA, 1994.
2. J. R. Patel, A. M. Shekh, and H. C. Ratanpara, "Seasonal incidence and effect of minimum temperature and vapour pressure on the population of mango hoppers in middle Gujarat," Gujrat Agricultural University Research Journal, vol. 20, pp. 5–8, 2004.
3. R. B. Fishwick, "Pesticide residues in grain arising from post-harvest treatments," Aspects of Applied Biology, vol. 17, no. 2, pp. 37–46, 1988.
4. M. Jacobson, R. E. Redfern, and G. D. Mills Jr., "Naturally occurring insect growth regulators. II. Screening of insect and

plant extracts as insect juvenile hormone mimics," Lioydia, vol. 38, no. 6, pp. 455–472, 1974.

5. G. Singh and S. K. Kaushik, "Comparative efficacy and sampling techniques for jassid population estimation on okra," Indian Journal of Ecology, vol. 17, pp. 58–60, 1990.
6. C. F. Henderson and E. W. Tilton, "Tests with acaricides against the brown wheat mite," Journal of Economic Entomology, vol. 48, pp. 157–161, 1955.
7. H. P. Misra and B. Senapati, "Evaluation of new insecticides against okra jassid (Amrasca biguttula),"Indian Journal of Agricultural Sciences, vol. 73, no. 10, pp. 576–578, 2003.
8. G. Santharam, K. Kumar, S. Chandrasekaran, and S. Kuttalam, "Bio efficacy and residues of imidacloprid in chillies used against chilli thrips," Madras Agricultural Journal, vol. 90, pp. 395–399, 2003.
9. F. Akbar, N. Yasmin, M. F. Khan, F. Naz, and T. A. Latif, "Efficacy of imidacloprid and endosulfan in comparison with biosal (Biopesticide) against Amrasca devastans Dist. on mustard," Pakistan Journal of Entomology, vol. 23, pp. 27–30, 2008.
10. N. C. Patel, J. J. Patel, D. B. Jayani, J. R. Patel, and B. D. Patel, "Bio efficacy of conventional insecticides against pests of okra," Indian Journal of Entomology, vol. 59, pp. 51–53, 2003.
11. M. F. Akbar, N. Yasmin, M. F. Khan, F. Naz, and T. A. Latif, "Effectiveness of profenophos in comparison with endosulfan against Amrasca devastans Dist. on okra crop," Pakistan Journal of Entomology, vol. 21, pp. 25–27, 2006.
12. M. F. Akbar, N. Yasmin, M. F. Khan, F. Naz, andT. A. Latif, "Relative efficacy of biopesticide in comparison with conventional pesticides against Amrasca devastans Dist. on brinjal," Pakistan Journal of Entomology, vol. 22, pp. 1–3, 2007.
13. K. Narottam, "Evaluation of some new insecticides and bio pesticides against jassid, (Amrasca biguttula biguttula) on okra, Abelmoschus esculentus," Indian Journalof applied Entomology, vol. 20, p. 1, 2006.
14. S. N. H. Naqvi, "Prospects and development of a neem based pesticide in Pakistan," Proceedings Pakistan Congress Zoology, vol. 16, pp. 325–338, 1996.

15. H. Schmutterer, "Properties and potential of natural pesticides from the neem tree, Azadirachta indica,"Annual Review of Entomology, vol. 35, pp. 271–297, 2000.
16. M. Grainge and S. Ahmad, Handbook of Plants with Pest Control Properties, John Wiley & Sons, New York, NY, USA, 1988.
17. A. Adlakshmi, M. D. Korat, and P. R. Vaishnav, "Effect of organic manures and inorganic fertilizers on the incidence of insect pests infesting okra," Karnataka Journal of Agricultural Science, vol. 21, pp. 287–289, 2008.
18. S. Dhingra, S. Walia, J. Kumar, S. Singh, G. Singh, and B. S. Parmar, "Field efficacy of azadirachtin-A, tetrahydroazadirachtin-A, NeemAzal and endosulfan against key pests of okra (Abelmoschus esculentus)," Pest Management Science, vol. 64, no. 11, pp. 1187–1194, 2008. · ·
19. M. F. Akbar, N. Yasmin, M. F. Khan, F. Naz, and T. A. Latif, "Effectiveness of different spray schedules against population of whitefly, Bemisia tabaci (Genn.) on okra crop," Pakistan Journal of Entomology, vol. 24, pp. 45–48, 2009.
20. S. N. H. Naqvi, S. M. Norulain, M. A. Azmi, and S. T. Asdaque, "Effect of neem fractions and malathion against white flies Aleurobus barodensis on Brinjal crop (Solanum melongena)," Sarhad Journal of Agriculture, vol. 5, pp. 25–28, 1989.
21. T. Basedow, H. R. Ossiewatsch, J. A. Bernal Vega, S. Kollman, H. A. F. El Shafie, and C. M. Y. Nicol, "Control of aphids and whiteflies (Homoptera: Aphididae and Aleyrodidae) with different Neem preparations in laboratory, greenhouse and field: effects and limitations," Journal of plant Disease and Protection, vol. 109, no. 6, pp. 612–623, 2002.
22. K. B. Pun, D. Sabitha, and R. Jeyarajan, "Management of okra yellow vein mosaic virus disease and its whitefly vector," Indian Journal of Virology, vol. 16, pp. 32–35, 2005.
23. M. Aslam and S. N. H. Naqvi, "The efficacy of a phytopesticide in comparison with perfekthion against sucking pests of cotton," Turkish Journal of Zoology, vol. 24, no. 4, pp. 403–408, 2000.
24. M. K. Khattak, Mamoon-ur-Rashid, A. S. Hussain, and T. Islam, "Comparative effect of neem (Azadirachtin indica A. Juss) oil, neem seed water extract and Baythroid TM against whitefly,

jassid and thrips on cotton," Pakistan Journal of Entomology, vol. 28, pp. 31–37, 2006.

25. P. Misrah, "Field evaluation of some newer insecticides against aphids (Aphis gossypii) and jassids (Amrasca biguttula) on okra," Indian Journal of Entomology, vol. 64, pp. 80–84, 2004.

Chapter 7

Chemical Composition and Acaricidal Effects of Essential Oils of *Foeniculum vulgare* Mill. (Apiales: Apiaceae) and *Lavandula angustifolia* Miller (Lamiales: Lamiaceae) against *Tetranychus urticae* Koch (Acari: Tetranychidae)

Asgar Ebadollahi[1], Jalal Jalali Sendi[1], Alireza Aliakbar[2], and Jabraeil Razmjou[3]

[1]Department of Plant Protection, Faculty of Agricultural Sciences, University of Guilan, Rasht 416351314, Iran

[2]Department of Chemistry, Faculty of Basic Sciences, University of Guilan, Rasht, Iran

[3]Department of Plant Protection, Faculty of Agriculture, University of Mohaghegh Ardabili, Ardabil, Iran

ABSTRACT

Utilization of synthetic acaricides causes negative side-effects on nontarget organisms and environment and most of the mite species such as two spotted spider mite, Tetranychus urticae Koch, are becoming resistant to these chemicals. In the present study, essential oils of fennel, Foeniculum vulgareMill., and lavender, Lavandula angustifolia Miller, were hydrodistilled using Clevenger apparatus and chemical composition of these oils was analyzed by GC-MS. Anethole (46.73%), limonene (13.65%), and -fenchone (8.27%) in the fennel essential oil and linalool (28.63%), 1, 8-cineole (18.65%), and 1-borneol (15.94%) in the lavender essential oil were found as main components. Contact and fumigant toxicity of essential oils was assessed against adult females of T. urticae after 24 h exposure time. The essential oils revealed strong toxicity in both contact and fumigant bioassays and the activity dependeds on essential oil concentrations. Lethal concentration 50% for the population of mite (LC_{50}) was found as 0.557% (0.445–0.716) and 0.792% (0.598–1.091) in the contact toxicity and 1.876 μL/L air (1.786–1.982) and 1.971 μL/L air (1.628–2.478) in the fumigant toxicity for fennel and lavender oils, respectively. Results indicated that F. vulgare and L. angustifolia essential oils might be useful for managing of two spotted spider mite, T. urticae.

INTRODUCTION

Spider mites belong to the family Tetranychidae and are named because many members of this family produce silk webbing on the host plants. Some 1,200 species of spider mites belonging to over 70 genera are known in the world especially in the Southern Hemisphere [1]. Two-spotted spider mite,Tetranychus urticae Koch, is widely distributed globally and a common pest of many plant species in greenhouses, orchards and field crops. To date, 3877 host species have been reported around the world in both outdoor crops and greenhouses [2]. T. urticae feeding causes graying or yellowing of the leaves and necrotic spots occur in advanced stages of leaf damage. Mite damage to the open flower causes a browning and withering of the petals that resembles spray burn. In addition, small chlorate spots can be formed at feeding sites as the mesophyll tissue collapses due to the destruction of 18–22 cells per minute [2]. The importance of this mite pest is not only due to direct damage to plants but also due to indirect damage to plants which decreases photosynthesis and transpiration [3]. Because of their high reproductive rates, its management can be difficult. When mites begin to feed on a plant, they produce webbing that can protect both motile and egg stages from the acaricide [3].

Synthetic acaricides have been used as the main strategy for Tetranychusspecies resulting in an increased cost for production and environmental impacts as well as resistance development to even the newly synthesized molecules such abamectin [2, 4]. In addition, control methods based on the use of synthetic acaricides sometimes fail to keep the number of spider mites below economic threshold levels [5]. It is therefore necessary to find alternatives that can minimize negative effects of synthetic acaricides.

Essential oils obtained by hydrodistillation, steam distillation, dry distillation, or mechanical cold pressing of aromatic plants [6] have long been used as fragrances and flavorings in the perfume and food industries, respectively, and recently for aromatherapy and medicines [2]. The essential oils can play major roles in pollination by attracting insects and in water loss's prevention due to excessive evaporation. Repellence is another property of essential oils, as some contain numerous secondary metabolites that can deter attacks from pests [7]. Most essential oil constituents degrade quickly in the environment or are rapidly lost from plant foliage through volatilization, which minimizes residual contact. They have short residual activities due to temperature and UV light degradation and, with a few exceptions, their mammalian toxicity is low [8, 9]. Therefore, essential oils can be applied to both field and greenhouse crops in the same manner as current synthetic acaricides [2, 9].

The fennel, Foeniculum vulgare Mill. [Apiales: Apiaceae (Umbelliferae)], is indigenous to the Mediterranean and is largely used to impart flavor to a number of foods, such as soups, sauces, pickles, breads, and cakes. It is an annual, biennial, or perennial herbaceous plant, depending on the variety, which grows in good soils from sunny mild climatic regions and is a well-known aromatic plant species. Traditionally in Europe and Mediterranean areas, fennel is used as antispasmodic, diuretic, anti-inflammatory, analgesic, secretolytic, galactagogue, eye lotion, and antioxidant remedy [10, 11]. The lavender, Lavandula angustifolia Miller [Lamiales: Lamiaceae (Labiatae)], is an evergreen bushy shrub with straight, woody branches; the lower parts are leafless, putting out numerous herbaceous stems to a height of about 1 m [12]. It is native to southern Europe and the Mediterranean area and is commercially cultivated in France, Spain, Portugal, Hungary, UK, Bulgaria, Australia, China, and USA [13].

This paper describes a laboratory study examining the contact and fumigant toxicity of essential oils of F. vulgare and L. angustifolia grown in Iran against ofT. urticae followed by evaluation of their chemical constituents by Gas chromatography-Mass spectrometry (GC-MS).

MATERIALS AND METHODS

Rearing of Two-Spotted Spider Mite

The two-spotted spider mites were collected from infested leaves of some wildly grown weeds in the yard of University of Mohaghegh Ardabili which did not have any exposure to acaricides. The Tetranychus urticae Koch species after separation, and slide preparation was identified according to introduced keys by Zhang [1]. Spider mites were reared on navy bean (Vigna unguiculata Walp. [Fabales: Fabaceae]) Plants for one year. The infested plants were held in cages (120 × 300 × 100 cm) covered with mesh cloth. To synchronize the adult stage of T. urticae for adulticidal bioassays, 50 adult female mites were transferred to the leaves of trifoliate bean plants (held individually in cylindrical glass containers with appropriate aeration) with a hair brush and allowed to lay eggs for 24 h, after which the adults were removed. The infested leaves were held at the above-mentioned conditions to allow the eggs to hatch and the larvae to develop into synchronized adults. All experiments were carried out at 25 ± 2°C, 60 ± 5% relative humidity (RH) and a photoperiod of 16: 8 (light : dark) in a growth chamber.

Plant Materials and Essential Oil Extraction

Aerial parts 5 cm from the top of L. angustifolia at flowering stage and seeds ofF. vulgare were collected from Ardabil, Ardabil province, Iran from June to August 2013. The specimens were air dried in the shade at room temperature and chopped into small pieces with electric grinder. The essential oils were extracted using a Clevenger-type water steam distillation apparatus within 3 h. To carry out the extraction, 100 g of powdered plant material was used along with 1200 mL distillated water. Anhydrous sodium sulphate was used to remove excess water after extraction. The essential oils were transferred to dark brown glass vials covered with aluminum foil and stored in refrigerator at 4°C until used in the experiments.

Analysis of Essential Oil

One μL of prepared essential oil was injected to GC-MS (HP Agilent 6800N/(61530N) with CPSil5CB column (Chrompack, 100% dimethyl polysiloxane 60 m, 0.25 mm (ID) film thickness 0.25 l). The analysis was performed under temperature programming from 100°C (3 min) to 250°C (5 min) with the rate of 3°C/min. Injector temperature was 230°C. Identification of spectra was carried out by study of their fragmentation and also by comparison with standard spectra present in the library of the instrument. Area normalization was used for determination of composition percentage.

Contact Toxicity

Contact toxicity was conducted in Petri dishes (6 cm diameter). Concentrations ranged from 0.12% to 2.8% for L. angustifolia and 0.11% to 1.7% for F. vulgare, using a spreader sticker adjuvant (20 μL Tween, 0.02%) diluted in distilled water. Leaf discs (3 cm diameter) were cut from leaves of greenhouse-grown V. unguiculata and immersed in solutions of the each essential oil for 20 seconds. After drying at room temperature for 45 min, each disc was individually placed at the bottom of a Petri dish atop a 6 cm diameter disc of filter paper wetted with distilled water. The wet cotton pads were then placed on excised leaves and ten adult females were transferred thereafter. The lids of

Petri dishes were pierced (1 cm in diameter) and their openings were covered with mesh cloth in order to evade of fumigant toxicity. Control mites were held on leaf discs immersed in dilutions without essential oils. Mortality was noted after 24 h and there were tree replicates for each treatment. Mites were considered dead if appendages did not move when prodded with a fine paintbrush.

Fumigant Toxicity

For evaluation of fumigant toxicity, 750 mL plastic containers with tight lids were used as the test chambers. Each treatment consisted of five concentrations of the essential oil and a control. Based on preliminary experiments, ranges of concentrations tested against the adult females were 0.53 to 4.2 and 1.33 to 2.4 μL/L air for L. angustifolia and F. vulgare, respectively. For each concentration, three replicates were used. Each replicate consisted of sixty 24 h old adult females on each leaf plant disc. The discs (3 cm in diameter) punched from leaves of bean plants were placed inside a 6 cm diameter plastic Petri dishes without lids lined with water-soaked cotton. The Petri dishes were then put into plastic containers used as fumigation chambers. To achieve the desired concentration of the oil in the fumigation chambers, using a micropipette, the appropriate volume of the oil was applied on a 2 × 2 cm strip of Whatman no. 1 filter paper adhered to the inner surface of the fumigation chamber. The exposure period for assessing the adulticidal effect of the essential oils was 24 h. To determine mortality, the mites were touched with the tip of a fine hair brush. If the mite did not move, it was considered dead. The controls consisted of the same number of mites as the treatments; and were kept under the same conditions on leaf disks left untreated.

Analysis of Data

Experiments were arranged in a completely randomized design and data were analyzed by ANOVA. The mortality data were subjected to probit analysis using SPSS software to estimate LC_{50} values of the essential oils against T. urticae.

RESULTS

Using hydrodistillation process, fennel seed yielded 2.15% essential oil while lavender leaf yielded 2.01%. Results of analysis of the essential oils are presented in Table 1. Twenty six compounds were identified in the essential oil of fennel, representing 99.94% of the total essential oil sample while twenty five compounds were found in the lavender essential oil, representing 99.97% of the total essential oil sample. The major components were found to be anethole (46.73%), limonene (13.65%), -fenchone (8.27%), carvone (6.12%), and estragole (5.26%) for F. vulgare essential oil and Linalool (28.63%), 1, 8-Cineole (18.65%), 1-Borneol (15.94%), Camphor (8.20%), and Terpineol-4 (4.27%) forL. angustifolia essential oil. Amounts of monoterpenes hydrocarbon in the essential oils of F. vulgare and L. angustifolia were 20.43% and 8.55%, respectively. Monoterpenoids content of F. vulgare and L. angustifolia essential oils was 20.27% and 83.36% and the sesquiterpenes were 0.44% and 2.38%, respectively.

Table 1: Chemical analysis of essential oils of Foeniculum vulgare and Lavandula angustifolia grown in Iran by GC-MS

Compound	F. vulgare		L. angustifolia		Formula	Molecular weight (g/mol)	Classification
	RT	Percentage	RT	Percentage			
-Pinene	3.93	3.53	4.80	1.46	C10H16	136.23	Monoterpene
Camphene	4.17	0.32	5.08	0.68	C10H16	136.23	Monoterpene
Sabinene	4.51	0.67	5.54	0.96	C10H16	136.23	Monoterpene
-Pinene	4.61	0.20	5.64	2.16	C10H16	136.23	Monoterpene
Myrcene	4.74	1.09	5.84	1.39	C10H16	136.23	Monoterpene
-Phellandrene	5.05	0.54	—	—	C10H16	136.23	Monoterpene
Limonene	5.61	13.65	—	—	C10H16	136.23	Monoterpene
1-Octen-3-ol	—	—	5.73	0.46	C8H16O	128.21	Alcohol
-Terpinene	6.06	0.43	—	—	C10H16	136.23	Monoterpene
-3-Carene	—	—	6.27	0.72	C10H16	136.23	Monoterpene
-Fenchone	6.83	8.27	—	—	C10H16O	152.23	Monoterpenoid
1,8-Cineole	—	—	6.87	18.65	C10H18O	154.25	Monoterpenoid
trans- -Ocimene	—	—	7.10	0.98	C10H16	136.23	Monoterpene
Terpineol-4	—	—	7.58	4.27	C10H18O	154.25	Monoterpenoid
Terpinolene	—	—	7.99	0.50	C10H16	136.23	Monoterpene
Linalool	—	—	8.64	28.63	C10H18O	154.25	Monoterpenoid
cis-Sabinenehydrate	—	—	8.99	0.51	C10H18O	154.25	Monoterpenoid
Camphor	7.92	0.28	9.58	8.20	C10H16O	152.23	Monoterpenoid
Estragole	9.12	5.26	—	—	C10H12O	148.20	Aromatic hydrocarbon

trans-Dihydrocarvone	9.24	0.47	—	—	C10H16O	152.23	Monoterpenoid
D-Fenchyl alcohol	9.49	0.18	—	—	C10H18O	154.25	Monoterpenoid
Fenchol	9.83	1.78	—	—	C10H18O	154.25	Monoterpenoid
1-Borneol	—	—	10.17	15.94	C10H18O	154.25	Monoterpenoid
Carvone	10.18	6.12	—	—	C10H14O	150.22	Monoterpenoid
-Carvone	10.26	3.17	—	—	C10H14O	150.22	Monoterpenoid
Hexyl butyrate	—	—	10.43	1.56	C10H20O2	172.26	Fatty ester
Cryptone	—	—	10.51	0.58	C9H14O	138.21	Ketonic chelate
-Terpineol	—	—	10.61	3.25	C10H18O	154.25	Monoterpenoid
Anethole	11.23	46.73	—	—	C10H12O	148.21	Aromatic hydrocarbon
Bornyl formate	—	—	11.29	0.61	C11H18O2	182.26	Ethyl ester
Linalyl acetate	—	—	11.83	2.18	C12H20O2	196.29	Monoterpenoid
Geraniol acetate	—	—	12.55	1.05	C12H20O2	196.29	Monoterpenoid
Eugenol	12.72	3.75	—	—	C10H12O2	164.20	Carboxylic Acid
Thymol	—	—	12.80	0.68	C10H14O	150.28	Monoterpenoid
cis-Jasmone	13.09	0.69	—	—	C11H16O	164.24	Fatty Acid
Anisyl acetone	13.17	0.36	—	—	C13H16O2	204.27	Ketonic ether
-Caryophyllene	13.90	0.22	—	—	C15H24	204.35	Sesquiterpene
Germacrene d	15.11	0.22	—	—	C15H24	204.35	Sesquiterpene
Eugenyl acetate	15.96	0.91	—	—	C12H14O3	206.24	Carboxylic Acid
-Farnesene	—	—	16.05	0.81	C15H24	204.35	Sesquiterpene
Benzeneacetic acid	17.07	0.17	—	—	C8H8O2	136.15	Carboxylic Acid
cis-isoapiole	17.85	0.28	—	—	C12H14O4	222.24	Phenylpropanoids

-Bisabolol	—	—	20.56	1.57	C15H26O	222.37	Sesquiterpenoid
-ethyl-4,4-dimethoxy-Stilbene	30.91	0.64	—	—	C16H16O2	240.30	Ketonic chelate
1,2-Benzenedicarboxylic acid	—	—	33.67	2.17	C8H6O4	166.13	Carboxylic Acid
Monoterpene hydrocarbons		20.43		8.85			
Oxygenated monoterpenes		20.27		83.36			
Sesquiterpene hydrocarbons		0.44		0.81			
Oxygenated sesquiterpenes		0		1.57			
Others		58.8		5.38			
Total		99.94		99.97			
Yield		2.15		2.01			

RT: retention time (min).

F. vulgare and L. angustifolia essential oils revealed strong significant toxicity on the adult females of T. urticae in both contact and fumigant assays. The activity depended on essential oil concentrations in both contact and fumigant bioassays and increased susceptibility of mite was directly associated with oil concentration (Table 2 and Figure 1).

Table 2: Contact and fumigant toxicity of the essential oils isolated from Foeniculum vulgare and Lavandula angustifolia against the adult females of Tetranychus urticae

Bioassay	Essential oil	Results of ANOVA		Results of probit analysis				Toxicity index
		F(df = 4, 10)a	P value	24-h LC50 with 95% confidence limitsb	Slope ± SE	χ^2 (df = 3)	Sig. c	
Contact toxicity	F. vulgare	49 .75	1. 4446	0.557 (0.445–0.716)	1.181 ± 0.144	3. 327	0. 351	100.00
	L. angustif olia	33 .30	9. 3842	0.792 (0.598–1.091)	0.936 ± 0.124	1. 620	0. 655	0.703
Fumigant toxicity	F. vulgare	37 .70	0. 00 005	1.876 (1.786–1.982)	5.335 ± 0.670	3. 086	0. 379	100.00
	L. angustif olia	18 .167	0.0 001	1.971 (1.628–2.478)	1.377 ± 0.187	3. 975	0. 264	0.951

[a]Calculated values are greater than values in table (= 0.05, credit = 3.4780). Therefore, they are significant.

[b]% v/v and μL/L air for contact and fumigant toxicity, respectively.

[c]Since the significance level is greater than 0.150, no heterogeneity factor is used in the calculation of confidence limits.

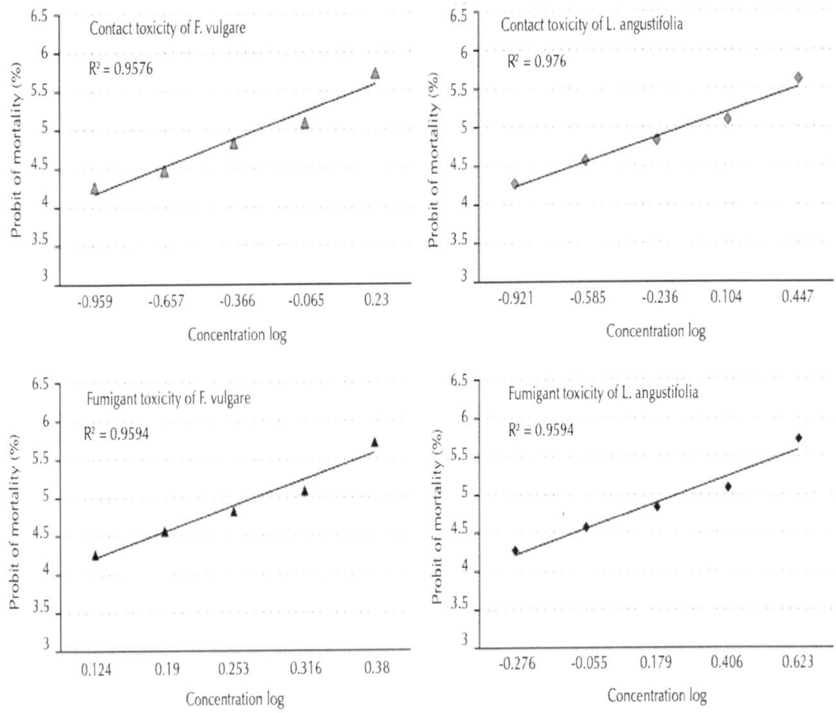

Figure 1: Concentration-mortality response lines for adult females of Tetranychus urticae exposed to different concentrations of Foeniculum vulgare and Lavandula angustifolia essential oils.

In the contact toxicity, lethal concentration 50% mite mortality (LC_{50}) was 0.557% and 0.792% with F. vulgare and L. angustifolia, respectively, with F. vulgare essential oil being the most toxic the adult females of T. urticae (Table 2). On the other hand, fumigant toxicity was 1.876 and 1.971 μL/L air for F. vulgare and L. angustifolia essential oils, respectively (Table 2).

DISCUSSION

In the present study, anethole, limonene, α-fenchone, and carvone were the major compounds of essential oil of F. vulgare while linalool, 1, 8-cineole, 1-borneol and camphor were the main compounds of essential oil of L. angustifolia. In the study of Chowdhury et al. [14], anethole (58.5% in seed oil and 51.1% in leaf oil) and limonene (22.9%

in leaf oil and 19.6% in seed oil) determined as main components of the seeds and leaves of F. vulgare. In the other study, borneol, α-terpinene, linolool, and geranyl proprionate were found as major constituents in the L. angustifolia essential oil. The variations could be due to differences in location, elevation, and genetic makeup of the plant or due to an adaptive process to particular ecological conditions [15].

Acaricidal activity of F. vulgare and L. angustifolia essential oils reported in the present study has been reported earlier by other authors. For example, the essential oils from F. vulgare and L. angustifolia were toxic against Varroa destructor Anderson and Trueman (a major pest of honey bees, Apis melliferaL.) [16]. The fumigant toxicity of essential oil extracted from seeds of F. vulgarewas tested against adult females of T. urticae by Amizadeh et al. [17].

Essential oils are characterized by two or three major components at fairly high concentrations (20–70%) compared to others components present in trace amounts. The components include two groups of distinct biosynthetical origin. The main group is composed of terpenes and terpenoids and the other of aromatic and aliphatic constituents, all characterized by low molecular weight [6]. They are made from combinations of several 5-carbon-base (C_5) units called isoprene. The monoterpenes are formed from the coupling of two isoprene units (C_{10}). They are the most representative molecules of the essential oils and allow a great variety of structures. A terpene containing oxygen is called a terpenoid. Hence, monoterpenes are found as two forms; monoterpenes hydrocarbon and oxygenated monoterpenes or monoterpenoids. The sesquiterpenes are formed from the assembly of three isoprene units (C_{15}) [6]. Monoterpenes hydrocarbon, monoterpenoids, and sesquiterpenes are present in our tested study too. Regarding their biological properties, essential oils are complex mixtures of numerous molecules, and their biological effects are the result of a synergism of all components or reflect only those of the main components present at the highest levels according to gas chromatographical analysis [6, 18]. It is suggested that the variability of biological activities of essential oils extracted from different plant species against T. urticae could be due to chemical components, differences in their chemical composition, and even in synergic and antagonistic interactions between these components [2, 19, 20].

One of the most attractive features of essential oils is that they are low-risk products and they are relatively well-studied experimentally and clinically because of their use as medicinal products [11]. In terms of ecotoxicology, essential oils are safe to use but not without potential problems. For example, constituents of essential oils are biodegradable, with short half-lives ranging from 30 to 40 h for α-terpineol [21] and in contrast to some synthetic insecticides; no biomagnification has been reported to date [11]. Their short residual half-lives on plants also enhance their compatibility with biological control agents and indigenous natural enemies of pests and reduce risks to honeybees and other foraging pollinators [2]. As cost effective commercial problems, large quantities of plant material must be processed to obtain sufficient quantities of essential oils for commercial-scale tests, situation which also requires breeding these plants in great quantities.

REFERENCES

1. Z. Q. Zhang, Mites of Greenhouses; Identification, Biology and Control, CABI Publishing, Wallingford, UK, 2003.
2. M. B. Isman, S. Miresmailli, and C. MacHial, "Commercial opportunities for pesticides based on plant essential oils in agriculture, industry and consumer products," Phytochemistry Reviews, vol. 10, no. 2, pp. 197–204, 2011
3. R. L. Brandenburg and G. G. Kennedy, "Ecological and agricultural considerations in the management of two spotted spider mite (Tetranychus urticae Koch)," Agricultural Zoology Reviews, vol. 2, pp. 185–236, 1987.
4. T. Ramasubramanian, K. Ramaraju, and A. Regupathy, "Acaricide resistance in Tetranychus urticae Koch (Acari: Tetranychidae)—global scenario," Journal of Entomology, vol. 2, pp. 33–39, 2005.
5. P. Tirello, A. Pozzebon, S. Cassanelli, T. van Leeuwen, and C. Duso, "Resistance to acaricides in Italian strains of Tetranychus urticae: toxicological and enzymatic assays," Experimental and Applied Acarology, vol. 57, no. 1, pp. 53–64, 2012.
6. F. Bakkali, S. Averbeck, D. Averbeck, and M. Idaomar, "Biological effects of essential oils—a review," Food and Chemical Toxicology, vol. 46, no. 2, pp. 446–475, 2008.

7. M. B. Isman, "Botanical insecticides, deterrents, and repellents in modern agriculture and an increasingly regulated world," Annual Review of Entomology, vol. 51, pp. 45–66, 2006.
8. S. Miresmailli and M. B. Isman, "Efficacy and persistence of rosemary oil as an acaricide against twospotted spider mite (Acari: Tetranychidae) on greenhouse tomato," Journal of Economic Entomology, vol. 99, no. 6, pp. 2015–2023, 2006.
9. C. Regnault-Roger, C. Vincent, and J. T. Arnason, "Essential oils in insect control: low-risk products in a high-stakes world," Annual Review of Entomology, vol. 57, pp. 405–425, 2012
10. L. S. Moura, R. N. Carvalho Jr., M. B. Stefanini, L. C. Ming, and M. A. A. Meireles, "Supercritical fluid extraction from fennel (Foeniculum vulgare): global yield, composition and kinetic data," Journal of Supercritical Fluids, vol. 35, no. 3, pp. 212–219, 2005
11. G. D. Semiz, A. Ünlükara, E. Yurtseven, D. L. Suarez, and I. Telci, "Salinity impact on yield, water use, mineral and essential oil content of fennel (Foeniculum vulgare Mill.)," Tarim Bilimleri Dergisi, vol. 18, no. 3, pp. 177–186, 2013. View at Scopus
12. M. Wichtl, Herbal Drugs and Phytopharmaceuticals, Medpharm Scientific Publishers, Stuttgart, Germany, 1994.
13. A. S. Shawl and S. Kumar, "Potential of lavender oil industry in Kashmir," Journal of Medicinal and Aromatic Plant Sciences, vol. 22, pp. 319–321, 2000.
14. J. U. Chowdhury, M. H. Mobarok, M. N. I. Bhuiyan, and N. C. Nandi, "Constituents of essential oils from leaves and seeds of Foeniculum vulgare Mill. cultivated in Bangladesh," Bangladesh Journal of Botany, vol. 38, no. 2, pp. 181–183, 2009.
15. V. A. Facundo, A. R. Polli, R. V. Rodrigues, J. S. L. T. Militao, R. G. Stabelli, and C. T. Cardoso, "Constituintes quımicos fixos e volateis dos talos e frutos de Piper tuberculatum Jacq. e das raızes de P. hispidum H. B. K," Acta Amazonica, vol. 38, pp. 733–742, 2008.
16. F. Kutukoglu, A. O. Girisgin, and L. Aydin, "Varroacidal efficacies of essential oils extracted from Lavandula officinalis, Foeniculum vulgare, and Laurus nobilis in naturally infested honeybee (Apis mellifera L.) Colonies," Turkish Journal of Veterinary and Animal

Sciences, vol. 36, no. 5, pp. 554–559, 2012

17. M. Amizadeh, M. J. Hejazi, and G. A. Saryazdi, "Fumigant toxicity of some essential oils on Tetranychus urticae (Acari: Tetranychidae),"International Journal of Acarology, vol. 39, no. 4, pp. 285–289, 2013.
18. S. Miresmailli, R. Bradbury, and M. B. Isman, "Comparative toxicity ofRosmarinus officinalis L. essential oil and blends of its major constituents against Tetranychus urticae Koch (Acari: Tetranychidae) on two different host plants," Pest Management Science, vol. 62, no. 4, pp. 366–371, 2006
19. C. Regnault-Roger and A. Hamraoui, "Fumigant toxic activity and reproductive inhibition induced by monoterpenes on Acanthoscelides obtectus (Say) (coleoptera), a bruchid of kidney bean (Phaseolus vulgarisL.)," Journal of Stored Products Research, vol. 31, no. 4, pp. 291–299, 1995
20. S. Attia, K. L. Grissa, G. Lognay, S. Heuskin, A. C. Mailleux, and T. Hance, "Chemical composition and acaricidal properties of Deverra scoparia essential oil (Araliales: Apiaceae) and blends of its major constituents against Tetranychus urticae (Acari: Tetranychidae)," Journal of Economic Entomology, vol. 104, no. 4, pp. 1220–1228, 2011.
21. G. Misra and S. G. Pavlostathis, "Biodegradation kinetics of monoterpenes in liquid and soil-slurry systems," Applied Microbiology and Biotechnology, vol. 47, no. 5, pp. 572–577, 1997

Chapter 8

Genomic Survey of the Ectoparasitic Mite Varroa Destructor, a Major Pest of the Honey Bee Apis Mellifera

R Scott Cornman[1], Michael C Schatz[2], J Spencer Johnston[3], Yan-Ping Chen[1], Jeff Pettis[1], Greg Hunt[1], Lanie Bourgeois[5], Chris Elsik[6], Denis Anderson[7], Christina M Grozinger[8], and Jay D Evans[1,4]

[1]USDA-ARS, Bee Research Laboratory, 10300 Baltimore Ave., Beltsville, MD 20705 USA

[2]Center for Bioinformatics and Computational Biology, University of Maryland, College Park, MD 20742 USA

[3]Department of Entomology, Texas A&M University, College Station, TX 77843 USA

[4]Department of Entomology, Purdue University, West Lafayette, IN 47907 USA

[5]USDA-ARS, Honey Bee Research Laboratory, 1157 Ben Hur Rd, Baton Rouge, LA 70820 USA

[6]Department of Biology, Georgetown University, 37th and O Streets, NW, Washington, DC 20057 USA

[7]CSIRO Entomology, Black Mountain Laboratories, Clunies Ross Street, Black Mountain ACT 2601, Australia

[8]Department of Entomology, Penn State University, University Park, PA 16802 USA

ABSTRACT

Background

The ectoparasitic mite *Varroa destructor* has emerged as the primary pest of domestic honey bees (*Apis mellifera*). Here we present an initial survey of the *V. destructor* genome carried out to advance our understanding of *Varroa* biology and to identify new avenues for mite control. This sequence survey provides immediate resources for molecular and population-genetic analyses of*Varroa-Apis* interactions and defines the challenges ahead for a comprehensive *Varroa* genome project.

Results

The genome size was estimated by flow cytometry to be 565 Mbp, larger than most sequenced insects but modest relative to some other Acari. Genomic DNA pooled from ~1,000 mites was sequenced to 4.3× coverage with 454 pyrosequencing. The 2.4 Gbp of sequencing reads were assembled into 184,094 contigs with an N50 of 2,262 bp, totaling 294 Mbp of sequence after filtering. Genic sequences with homology to other eukaryotic genomes were identified on 13,031 of these contigs, totaling 31.3 Mbp. Alignment of protein sequence blocks conserved among *V. destructor* and four other arthropod genomes indicated a higher level of sequence divergence within this mite lineage relative to the tick *Ixodes scapularis*. A number of microbes potentially associated with *V. destructor* were identified in the sequence survey, including ~300 Kbp of sequence deriving from one or more bacterial species of the Actinomycetales. The presence of this bacterium was confirmed in individual mites by PCR assay, but varied significantly by age and sex of mites. Fragments of a novel virus related to the Baculoviridae were also identified in the survey. The rate of single nucleotide polymorphisms (SNPs) in the pooled mites was estimated to be 6.2×10^{-5} per bp, a low rate consistent with the historical demography and life history of the species.

Conclusions

This survey has provided general tools for the research community and novel directions for investigating the biology and control of *Varroa* mites. Ongoing development of *Varroa* genomic resources will be a boon for comparative genomics of under-represented arthropods, and will further enhance the honey bee and its associated pathogens as a model system for studying host-pathogen interactions.

BACKGROUND

Honey bees (*Apis mellifera*) are an important agricultural commodity providing honey, other bee products, and pollination services [1, 2]. Domesticated honey bees in the United States and elsewhere have been in decline in recent years, despite an increasing need for honey bee pollination services [3]. This fact is often blamed on increasing challenges from pests and pathogens, as well as episodes of severe decline such as the enigmatic 'colony collapse disorder' (CCD) [4].

Among the most detrimental of honey bee pests is the ectoparasitic mite *Varroa destructor* [5]. *V. destructor* and its closely related congener, *V. jacobsoni*, are native to Asia where they parasitize the Eastern honey bee, *A. cerana*. *V. destructor* was only identified as a morphologically and genetically distinct species from *V. jacobsoni* relatively recently [6]. *V. destructor* began to appear in Asian colonies of *A. mellifera* during the last century and is now widely distributed, inadvertently aided by trade in bees and bee products.

Mite-infested bee colonies suffer directly from parasitism of pupae and adults, and indirectly from viral and microbial pathogens that the mites vector [7, 8]. Feeding by mites induces an immunosupression in bees that leads to increased titres of pre-existing infections [9], further compounding their impact. The economic toll of *V. destructor* on apiculture is estimated to be millions of U.S. dollars per year, and chemical control agents are worrisome both for their collateral effects on bee health and the potential for honey contamination [10].

Varroa-honey bee interactions are mediated to a large extent via chemical cues, and bees have numerous mechanisms to control *Varroa* populations (reviewed in [5, 11]). *Varroa* mites reproduce on honey bee pupae, using chemical signals produced by the developing honey bee larvae to target appropriately aged hosts. The mature female offspring of reproductive *Varroa* emerge with the adult honey bee, and subsequently move to nurse bees (which are engaged in brood care), thereby allowing them to remain in close proximity to the brood [12, 13]. Honey bees resist 'Varroatosis', the infestation of colonies by *Varroa* mites, via grooming of adult infested bees, removal of infested pupae (hygienic behavior), and physiological resistance mechanisms [5]. Recent successes in breeding *Varroa*-resistant bees, including the selection of 'Russian' bees with longstanding exposure to mites [14, 15], indicate that a better understanding of how bees and mites interact with each other can lead to novel management strategies.

Comparative studies of the fragility of the *A. mellifera - V. destructo* interaction, which has apparently prevented most Asian lineages of *V. destructo* as well as other *Varroa* species from colonizing *A. mellifera* [6,16-18], supports the hypothesis that mite olfaction or other requirements for mite reproduction may be suitable control targets. A molecular-genetic approach to develop such innovative controls would clearly benefit from further insights into *Varroa* genomics, which could be exploited in conjunction with tools already extant for honey bee. Prior to this study, genes for only two non-mitochondrial *V. destructor* proteins had been deposited in GenBank, a sodium channel gene (AAN37408.1) and a glycoprotein (ACU30143.1). Genome sequencing will greatly expand this gene catalog, and may also uncover unforeseen targets for novel and specific acaricides, such as divergence in metabolic pathways between mites and bees or the discovery of important microbial interactions.

High-throughput, shotgun sequencing of whole genomes allows the rapid identification of thousands of genic sequences, greatly facilitating molecular and population-genetic studies that would otherwise proceed in piecemeal and laborious fashion. Here we report an initial sequence survey of the *V. destructor* genome in conjunction with a flow-cytometric estimate of genome size. Our annotations and analysis should aid investigators seeking molecular approaches to mite control. They will also provide a guide for a planned full genome project for this species [19], one of several genomics initiatives that are unfolding the molecular interactions between honey bees and a constellation of potentially interacting pathogens [4,7,20,21].

Of the eight genetically distinct lineages of *V. destructo* that parasitize *A. cerana* in Asia, two have been identified on *A. mellifera* [6, 18, 22, 23]. Anderson [24] designated these lineages the Japan (J) and Korea (K) 'haplotypes' in reference to mitochondrial DNA makers, but they are concordantly distinct at nuclear markers as well [23]. Genetic differentiation within lineages is low [23], likely reflecting the population-genetic impact of life-history traits [5] such as full-sib mating and male haploidy [25], as well as potential population bottlenecks tied to host-shift events and subsequent range expansion [18,23]. In this study, we have analyzed the K haplotype of *V. destructor* from *A. mellifera*, the predominant haplotype presently found in North America [23]. We have identified over 13,000 contigs with sequences homologous to other species; many of these have recognized domains and/or functional annotations transferred from other arthropods. Interestingly, *V. destructor* appears to have experienced a higher rate of protein evolution than *Ixodes scapularis*since their divergence from the most recent common ancestor over 300 million years ago. Sequences attributable to a range of microbes were identified, including a large number of sequences from one or more novel actinomycete bacteria, the presence of which was confirmed by PCR in individual mites but not in adult honey bees. We also identified a novel virus related to the Baculoviridae that was abundant in the genomic survey. Finally, we found a low level of nucleotide polymorphism in the sequenced sample of ~1,000 mites, consistent with expectation [23]. This bodes well for future efforts to sequence and assemble a reference genome for this species and to identify genetic variation that correlates with host-interaction traits among *Varroa* strains and species.

RESULTS

Genome size

Flow cytometry of *V. destructor* nuclei (normalized to nuclei of *Drosophila virilis*) yielded a haploid genome size estimate of 565 ± 3 Mbp (Figure 1). This genome size is larger than that of many insects but substantially smaller than that of numerous mites and ticks for which genome projects are underway or have been proposed (> 2,000 Mbp, [26]. It is also lower than the general range for ticks, which spans from roughly 1,000 Mbp in the argasid soft tick *Ornithodoros turicata* to 3,100 Mb in the ixodid hard tick *Amblyomma americanum* [27]. However, examples of very small acarid genomes are known. Most notably, the two-spotted spider mite *Tetranychus urticae*, which has been advocated as a model for genetic and developmental studies of chelicerates [28]), has an estimated genome size of only 75 Mbp. It is not yet clear to what extent these large differences in genome size are driven by variation in gene content, repetitive fraction, and/or ploidy. While the contributions of the latter two factors have been frequently noted, the potential contribution of gene expansion has been highlighted by recent analyses of waterflea [29] and pea aphid [30] that reveal a roughly two-fold increase in gene content relative to other arthropods.

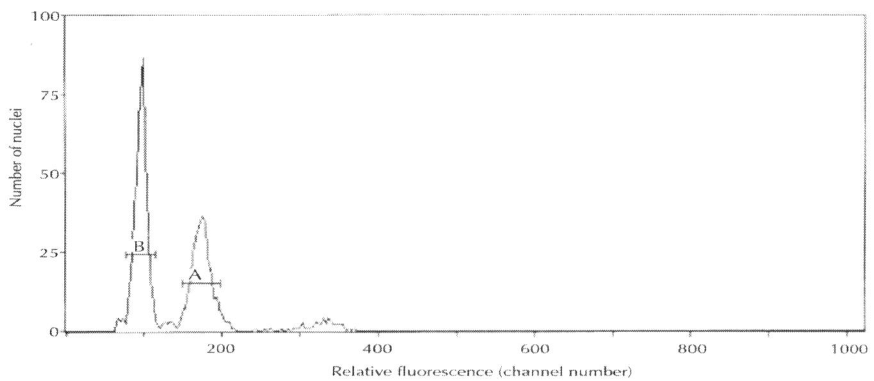

Figure 1: 2C genome size estimate for *Varroa destructor*. 2C genome size estimate for *V. destructor* based on flow cytometry, normalized to the*Drosophila*

virilis genome (A). The *V. destructor* fluorescence peak (B) corresponds to a genome size of 0.577 picograms or 565 ± 3 Mbp.

Sequencing, Assembly, and Filtering

Six pyrosequencing runs on a Genome Sequencer FLX instrument (454 Life Sciences) produced 2.4 Gbp of unpaired sequencing reads. After trimming low quality sequence, the average read length was 365 bp, generating an expected genome coverage of 4.2×. The reads were then assembled into contigs with the CABOG package of Celera Assembler version 5.2 [31]. The resulting assembly had a realized median contig coverage of 4.3×, but was highly fragmented and summed to only 318 Mbp of sequence, substantially less than the estimated genome size of 565 Mbp. Table 1 provides statistics for this assembly before and after removing problematic or undesired sequences (detailed below). The modest contig lengths are primarily a consequence of low coverage, as over 97% of contigs greater than 1 Kbp terminate due to coverage gaps rather than ambiguity from repetitive sequences. In fact, contig lengths were comparable to that expected (~3,860 bp) for an idealized assembly of equivalent coverage and read length, following the method of [32].

Table 1: Statistics of the *Varroa destructor* genome sequence survey

Statistic	Initial assembly	Filtered assembly
Number of contigs	271,543	184,094
Sum of contig length (Mbp)	318	294
Maximum contig length (bp)	18,703	16,332
Mean contig length (bp)	1,170.4	1,597.5
N50 contig length (bp)	2,107	2,262
Contigs ≥ 1,000 bp	107,195	105,621
Contigs ≥ 5,000 bp	5,407	5,374
Contigs ≥10,000 bp	120	118
Mean coverage	4.3	5.0

Cornman et al. BMC Genomics 2010 11:602, doi:10.1186/1471-2164-11-602

As detailed in the Methods, the small physical size of *V. destructor* required the collection of large numbers of mites from multiple honey bee colonies in order to obtain sufficient DNA for this survey. While mites were carefully cleaned and examined under a dissection microscope to remove any non-target organisms that might have been collected inadvertently, whole-organism extractions necessarily entail the possible inclusion of associated microbes, particularly gut microbes. The gut flora may include symbiotic, commensal, and pathogenic prokaryotes and eukaryotes, which are often important components of the ecology of arthropod species. We therefore filtered the assembled contigs based on G+C content, coverage, and sequence homology, in order to minimize the mis-annotation of microbial sequences as *V. destructor*, as well as to identify novel microbes of interest. Comparable strategies have been successfully applied to the classification of metagenomic samples (e.g., [33]).

We first examined the distribution of contig-mean coverage versus length (Figure 2), which revealed relatively few outliers, and these were found to be enriched in mitochondrial and ribosomal sequences. Thus, the vast majority of the assembled contigs do not appear to contain many collapsed nuclear repeats. Based on the observed median and variance of coverage (Figure 2), we excluded contigs with less than 2× coverage and greater than 10× coverage, as well as contigs less than 300 bp in length, from the analyzed *V. destructor* assembly.

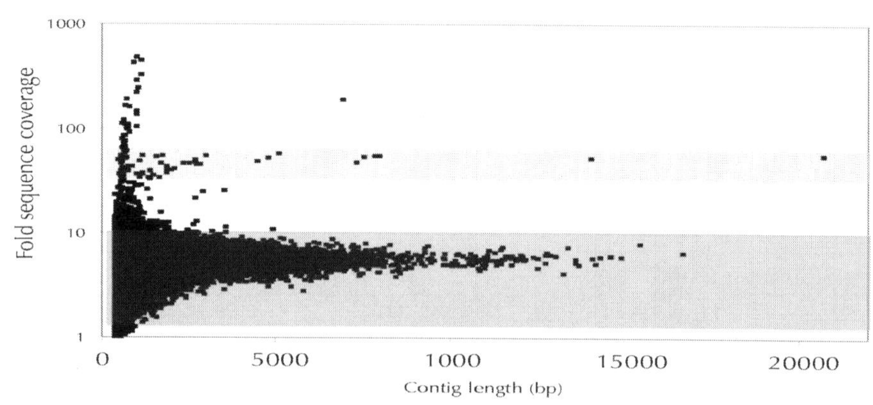

Figure 2: Scatterplot of contig coverage versus length for the unfiltered assembly. Fold sequence coverage of assembled contigs is distributed relatively narrowly around the mean, with few outliers. Contigs with coverage within the range 2-10× (shaded green) were included for further analysis as putative

nuclear genomic contigs of *V. destructor* (see text for details). Short, high-coverage contigs are predominantly mitochondrial or ribosomal in nature. A cluster of longer contigs approximately with 30-50× coverage appear to derive from a baculovirus (see text); contigs in this coverage range are shaded in yellow.

We then examined G+C content of contigs (Figure 3), which suggested that *Varroa* nuclear DNA falls largely between 32-58% G+C content (40.9% G+C on average). Contigs with lower G+C content and higher coverage showed strong homology to mitochondrial and ribosomal DNA sequences previously reported for *V. destructor,* as would be expected. Contigs with higher G+C consistently showed higher sequence similarity to bacterial sequences than to arthropod sequences. For example, Figure 4 illustrates the distribution of contigs with BLASTX matches to the high G+C bacterial order Actinomycetales at an expectation of 10^{-8}, plotted as a function of length and G+C content of contigs. Of these contigs, only those above approximately 58% G+C content were better matches (by sequence similarity and E-value) to Actinomycetales than to arthropods, or lacked an arthropod match entirely. Based on these considerations, we removed contigs outside the range of 32-58% G+C from the analyzed *V. destructor* assembly unless they contained a superior match to a eukaryotic sequence in GenBank at an expectation of 10^{-8}. Additional filtering was performed as described in the Methods to specifically remove sequences from organisms that were considered potential contaminants *a priori,* such as known microbial pathogens of honey bees that are dispersed as spores.

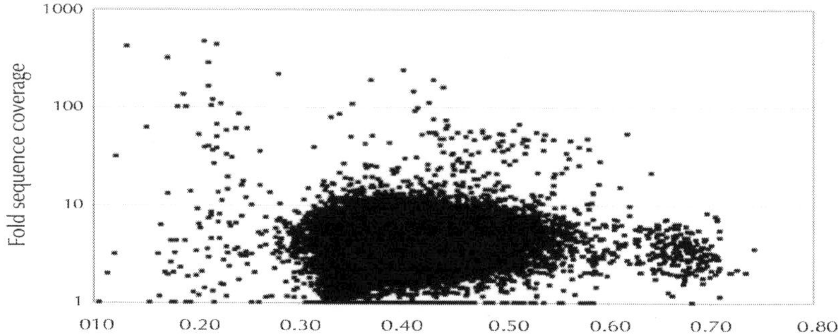

Figure 3: Scatterplot of contig G+C content versus coverage. Scatterplot of contig G+C content versus fold sequence coverage shows a clear mode of

G+C content in the range of 32-56%. The long tail of low G+C contigs includes low-complexity sequences such as AT repeats, as well as mitochondrial and ribosomal contigs. A secondary mode of high G+C contigs is also apparent; these contigs include many BLAST matches to the bacterial order Actinomycetales (see text).

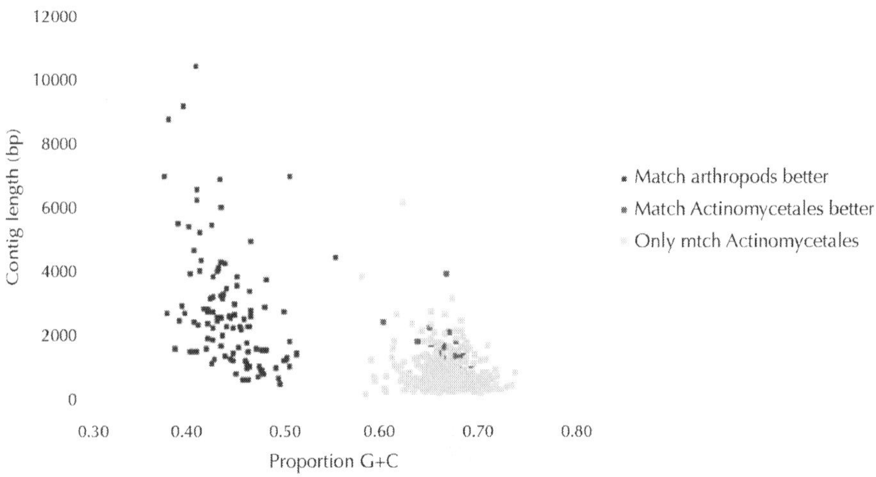

Figure 4: Evidence that high G+C contigs are bacterial. Contigs with a BLASTX match to the bacterial order Actinomycetales (at an expectation of 10^{-8} or less), plotted as a function of G+C content and contig length. Points are color-coded according to the taxonomy of the best GenBank match overall. There is a clean separation between contigs with lower G+C that are more similar, by percent identity and expectation of BLASTX sequence alignments, to arthropod sequences and contigs with higher G+C that are more similar to actinomycete sequences.

Actinomycete Discovery

The strong secondary mode of high G+C contigs in Figure 3 and taxonomically coherent BLAST hits (Figure 4) suggests that one or more Actinomycete species were particularly abundant in the sequenced sample. Together these contigs totaled ~300 Kbp of sequence, which were further analyzed with the BLAST2GO annotation tool [34]. includes BLAST2GO output that summarizes the distribution of BLASTX hits among these contigs with respect to matched species, expectation, and sequence similarity. All of the most frequently matched organisms are members of Actinomycetales. Ribosomal sequences were also found that had strong similarity to this clade, the closest match (98% identity) being to the genus *Segniliparus*.

To further investigate the distribution of this bacterium in *Varroa* mites, we designed primers specific to a homolog of translation initiation factor TIF-3. These primers amplified the target sequence from DNA of individual mites, as confirmed by sequencing of the amplicon. The rate of infection in a sample of mites (see Methods) is shown in Table 2. Rates of infection varied significantly ($P < 0.01$, 2 test) by age and sex. Mature females were infected more often than males (61% versus 33%). We did not detect the infection in a small sample of eggs and it was rare in nymphs (11%). These data suggest that horizontal transmission of the bacteria occurs within capped cells (from which males do not leave). We were not able to amplify the target from *A. mellifera* DNA extracted from adult bee abdomens, either with these primers or another pair targeting an ABC transporter gene (see Methods). Given the intimacy of *Varroa* mites with their hosts, this result suggests that the bacterium has some specificity to *Varroa*. While additional surveys at broader geographical scales are needed to confirm and extend these conclusions, the amount of sequence classified as actinomycete is remarkable given that common microbial components of the arthropod gut flora [35] (e.g., the Enterobacteriaceae) were not strongly evident in the initial assembly (see Methods). However, as the mites were not surface sterilized prior to extraction of genomic DNA, it is unclear whether the infection is topical or internal.

Table 2: Frequency of infection of individual mites by a novel actinomycete bacterium identified in the *V. destructor* sequence survey

Class	Present	Absent
Male	4	8
Female	11	7
Nymph	2	16
Egg	0	5

Presence or absence of the bacterium is based on a PCR assay (see text for details). Cornman et al. BMC Genomics 2010 11:602, doi: 10.1186/1471-2164-11-602

Baculovirus Discovery

Curiously, the longest contig in the initial assembly was among those filtered due to high coverage across its entire length (a mean of 56×). In fact, this contig is part of a distinct group visible in Figure 2 that have mean coverage around 50×, sloping down to ~30× as contig length decreases (a frequently observed phenomenon attributable to low-coverage at contig ends). We therefore investigated whether contigs ≥ 1 Kbp in length and with 30-60× coverage were of a consistent type, such as microbial DNA or a class of repetitive sequence (e.g., ribosomal genes or transposable elements). BLASTX and Pfam [36] searches indicated that these high coverage contigs derive from a novel baculovirus. Baculoviruses are large DNA viruses that are common in certain arthropod taxa and often have strong impacts on host survival (reviewed by [37]). Known baculoviruses range from 80 to 180 kb in length and typically encode more than 100 ORFs, some of which are putatively specific to Baculoviridae. Indeed, several baculovirus-specific domains were detected among the 216 methionine-initiated ORFs greater than 90 amino-acids in length that were found on these high-coverage contigs (Table 3). Domains related to viral genome replication as well as other domains previously reported in Baculoviridae were also found. The baculovirus *Spodoptera litura* NPV appears to be the most similar viral species overall in GenBank, as there were strong matches to ribonucleotide reductase subunits RR1 and RR2 and weaker matches to a number of other peptides of this viral species. A number of ORFs encoding low-complexity proteins were identified from these high-coverage contigs, but it remains unclear which of these, if any, are viral in nature (all filtered contigs and predicted ORFs are given in PCR primers specific to two separate loci (see Methods) confirmed the presence of these sequences in the original sample pool, but we were able to amplify these loci from only two of fifty individual mites collected at different times from the same colonies as the sequenced sample. Thus, the putative baculovirus-like sequences do not appear to be common in *V. destructor* although we can infer from their high coverage in the initial assembly that high titres are sometimes achieved. This result also demonstrates that these sequences are not integrated into the *V. destructor* genome. Because baculoviruses are important tools for arthropod transgenics and biocontrol [38, 39], efforts to clarify the nature of this virus and its hosts are underway.

Table 3: Evidence for a novel virus related to the Baculoviridae in the sequenced sample of *Varroa destructor*

ORF	Pfam domain description	Expectation	Reported in Baculoviridae?
VDK 00007920-4466_1	Ribonucleotide reductase, barrel domain	5.50E-231	Yes
VDK 00121146-847_1	Ribonucleotide reductase, small chain	1.30E-125	Yes
VDK 00001240-6963_1	Thymidylate synthase	3.60E-114	Yes
VDK 00008686-4345_1	Kinesin motor domain	3.90E-060	No*
VDK 00103915-1040_1	Reverse transcriptase	7.80E-035	Yes
VDK 00064516-1660_1	BRO family, N-terminal domain	1.60E-017	Yes
VDK 00001041-7192_1	Chitin binding domain	1.50E-010	Yes
VDK 00192648-381_1	Pacifastin inhibitor (LCMII)	3.60E-008	No
VDK 00158309-530_2	Baculovirus hypothetical protein	1.10E-006	Yes
VDK 00025611-2890_1	Matrixin (matrix metalloprotease)	3.80E-006	Yes

VDK 00139482-672_1	Protein of unknown function (DUF666)	3.90E-006	Yes
VDK 00179897-428_1	Phosphatidylinositol-specific phospolipase	5.60E-006	No*
VDK 00008449-4382_2	Protein of unknown function (DUF686)	9.90E-006	Yes
VDK 00099267-1099_1	Zinc knuckle (retroviral gag protein)	1.20E-004	Yes
VDK 00107278-999_1	Baculovirus BRO family, N-terminal domain	1.80E-004	Yes
VDK 00158309-530_3	Gamma-glutamyltranspeptidase	2.60E-004	No*
VDK 00071395-1528_1	Collagen triple helix repeat	8.20E-004	No*
VDK 00038309-2357_1	Collagen triple helix repeat	9.20E-004	No*
VDK 00202991-345_3	Amelogenin (cell adhesion protein)	2.20E-003	No
VDK 00042225-2226_1	Alpha/beta hydrolase fold	2.40E-003	No*
VDK 00021332-3129_1	Phage integrase family	3.00E-003	No*
VDK 00104983-1027_1	Collagen triple helix repeat	4.20E-003	No*
VDK 00202991-345_2	Collagen triple helix repeat	4.20E-003	No*

VDK 00043167-2196_3	Protein of unknown function (DUF686)	9.10E-003	Yes
VDK 00073176-1494_3	Gamma-glutamyltranspeptidase	9.20E-003	No*
VDK 00048355-2046_1	Matrixin (matrix metalloprotease)	0.01	Yes

Pfam matches are shown for methionine-initiated ORFs of 90 amino acids or more that were encoded by contigs at least 1 kb in length and with 30×-60× coverage. Collectively the matches are consistent with a viral origin and include several domains characteristic of Baculoviridae. Only matches with an expectation of 0.01 or less are shown. *Domain reported from at least one virus in Pfam [36] and Interpro [66] databases. **Present in one baculovirus accession (Q65353 of UniProt [67]) due to a retrotransposon insertion.

Cornman et al. BMC Genomics 2010 11:602, doi:10.1186/1471-2164-11-602

V. Destructor Annotation and Evolutionary Comparison with Other Arthropods

We used BLASTX (for genomic contigs) and BLASTP (for ORFs of 90 residues or more) to identify genic sequences in the assembly. Sequences were initially searched against a database of five arthropod peptide predictions (*Drosophila melanogaster, Anopheles gambiae, Pediculus humanus* [a representative non-Dipteran insect], *Daphnia pulex*, and *I. scapularis*), with a minimum expectation of 10^{-8}, and then secondarily against the eukaryotic Refseq database at the same expectation. In total, 13,031 contigs were identified with BLAST-detected similarity to database sequences (listed in these contigs had a median length of 1,967 bp and summed to 31.3 Mbp, and represent roughly 10% of the total assembled sequence (294 Mbp after filtering). ORFs with significant Pfam domains

These annotated gene fragments are necessarily an incomplete accounting of the number and type of genes in *V. destructor*, given the limitations of the assembly and lack of transcriptome data. Additional genomic resources are needed for robust gene models and are in development [19]. However, we used two approaches to infer how well represented *Varroa* protein-coding genes are in this survey. We first identified ORFs that were putatively homologous to enzymes of the glycolysis/gluconeogenesis pathway, using KEGG-annotated pathway components [40] from the mosquito, *Anopheles gambiae*, and the tick, *I. scapularis*, as BLAST queries. *I. scapularis* is the most closely related organism to *V. destructor* for which an extensively annotated genome sequence exists [41]. We identified 19 putative pathway components in our search (Table 4), whereas there are 21 members in *I. scapularis* and 23 in *A. gambiae*. A second, similar approach to assessing how well *V. destructor* genes were represented in the assembly was to query all predicted ORFs against the CEGMA set of hidden Markov models of evolutionarily conserved proteins [42]. A Hmmer [43] search found matches for 303 of the 458 protein models available, at an E-value threshold of 1.0 (matches at this level also had BLASTP matches in GenBank with E-values less than 10^{-10}). The same search performed on *I. scapularis* predicted proteins found matches for 429 models. Given that ORFs are proxies for single coding exons, both approaches used here are likely to under-sample *V. destructor* coding sequences. These comparisons nonetheless suggest that a majority of the coding potential of *V. destructor* was captured in this survey, if in fragmented form. However, they also indicate that comparisons of gene-family or protein-domain abundance in *V. destructor*relative to other arthropods may be premature.

Table 4: The *Varroa destructor* glycolysis/gluconeogenesis pathway is well represented in the genome sequence survey

Glycolysis/gluconeogesis enzyme	KEGG ID	Annotated in A. gambiae	Annotated in I. scapularis	Closest contig match in the V. destructor assembly (BLASTX)	Strand	Start	Stop	E-value
6-phosphofructokinase	K00850	X	x	VDK00166959	+	274	486	6.00E-027
acetyl-CoA synthetase	K01895	X	x	VDK00052872	-	313	1554	3.00E-085
aldehyde dehydrogenase	K00128	X	x	VDK00013090	+	2385	3224	1.00E-138
dihydrolipoamide dehydrogenase	K00382	X	x	VDK00011534	-	343	917	1.00E-020
enolase	K01689	X	x	VDK00029529	-	752	2211	7.00E-143
fructose-1,6-bisphosphatase	K03841	X	x	VDK00132162	+	158	355	1.00E-023
fructose-bisphosphate aldolase	K01623	X	x	VDK00012888	+	801	2121	1.00E-073
glucose-6-phosphate isomerise	K01810	X	x	VDK00034893	+	2	145	6.00E-018
glyceraldehyde 3-phosphate dehydrogenase	K00134	X	x	VDK00020468	-	142	2477	2.00E-077
hexokinase	K00844	X	x	VDK00023511	-	313	1522	2.00E-094
phosphoenolpyruvate carboxykinase	K01596	X	x	VDK00063370	-	1166	1684	1.00E-038
phosphoglucomutase	K01835	X	x	VDK00033184	-	236	478	3.00E-024

phosphoglycerate kinase	K00927	X	x	VDK00052433	+	351	1229	4.00E-032
phosphoglycerate mutase	K01834	X	x	VDK00033074	-	257	2550	5.00E-031
pyruvate dehydrogenase E1 component, subunit alpha	K00161	X	x	VDK00079694	-	2	1382	8.00E-043
pyruvate dehydrogenase E1 component, subunit beta	K00162	X	x	VDK00041927	-	527	2110	2.00E-020
pyruvate dehydrogenase E2 component	K00627	X	x	VDK00062508	-	165	416	4.00E-033
pyruvate kinase	K00873	X	x	VDK00096436	-	676	1131	4.00E-067
S-(hydroxymethyl) glutathione dehydrogenase	K00121	X	x	VDK00003663	-	1731	2700	4.00E-125
glucose-6-phosphate 1-epimerase	K01792	X	x					
triosephosphate isomerise	K01803	X	x					
aldose 1-epimerase	K01785	X						
L-lactate dehydrogenase	K00016	X						

Putative pathway components listed below were identified by BLASTX, using KEGG-annotated components [40] from Anopheles gambiae and Ixodes scapularis as search queries.

Cornman et al. BMC Genomics 2010 11:602, doi:10.1186/1471-2164-11-602

We used RepeatMasker [44] to characterize the distribution of protein-coding transposable element classes in *V. destructor*, which are summarized in Table 5. The Mariner class of DNA transposon was by far the most abundant transposable element identified. Some retrotransposons were also common, particularly *gypsy*-type long-terminal-repeat (LTR) retrotransposons and LINEs (long interspersed nuclear elements). Helitrons, which replicate by a rolling-circle method, were also numerous. Given the quasi-clonal nature of the K and J haplotypes of *V. destructor* [23], comparative re-sequencing of these two groups could uncover recently active transposable elements. Active elements are of interest because they are important contributors to genome evolution, including the creation of novel host genes [45], and have utility for functional genomics[46].

Table 5: Protein-encoding transposable elements in *V. destructor*

Class	Family	Number identified
DNA	TcMar-Mariner	6511
DNA	TcMar-Tc1	357
DNA	hAT	58
DNA	TcMar-Fot1	27
DNA	MuDR	12
Helitron	Helitron	338
LINE	R1	981
LINE	L2	93
LINE	CR1	83
LINE	BovB	47
LINE	Jockey	28
LINE	L1	17
LTR	Gypsy	914
LTR	Pao	102
LTR	Copia	74
LTR	Gypsy-Cigr	31

Transposable elements identified with the protein-level repeat masking mode of RepeatMasker [44]. Class abbreviations are DNA, type II DNA transposon; LINE, long interspersed nuclear element; LTR, long-terminal repeat retrotransposon. Family designations are those of RepBase [68].

Cornman et al. BMC Genomics 2010 11:602, doi:10.1186/1471-2164-11-602

Although the *I. scapularis* genome should be a valuable guide for homology-based gene prediction in *V. destructor* these lineages are estimated to have diverged 336 ± 26 million years ago [47]. It is therefore of interest to assess the level of sequence divergence between these two taxa, as well as the divergence of these Acari from other model arthropods. We identified 730 peptide blocks, averaging 128 residues in length, that were conserved between putative orthologs in these two species and among their closest homologs in *Da. pulex, Dr. melanogaster*, and *P. humanus*. We then calculated the average genetic distance among species for each block, using the JTT substitution matrix [48] and weighted by block length (see Methods). The unrooted dendrogram derived from a total of 94,146 aligned positions is shown in Figure 5. The *V. destructor* branch (0.42333) is substantially longer than the *I. scapularis* branch (0.26667) from their shared common ancestor, implying a high rate of amino-acid evolution in the *Varroa* lineage and/or a low rate of amino-acid evolution in the *Ixodes* lineage. Note that this distance-based approach does not require specifying a nucleotide substitution model or correcting for multiple substitutions; the branch lengths are approximately proportional to the product of the time since divergence and the branch-specific rate of molecular evolution (see [49] for a discussion). Furthermore, the branching order of these taxa are well-supported by independent data [50, 51]. Of course, this result is averaged across many loci and does not necessarily represent the pattern of sequence divergence at any given gene. It remains to be seen whether other gene features, such as exon structure, have also evolved at a comparably divergent rates.

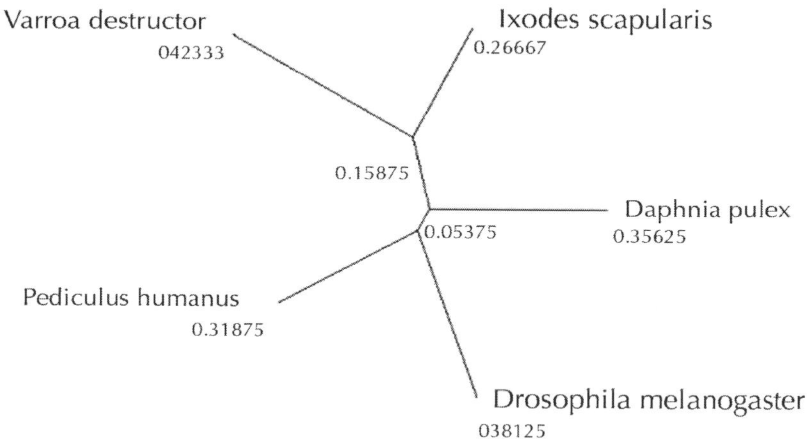

Figure 5: A distance-measure dendrogram of five arthropod taxa based on aligned blocks of conserved peptides. The unrooted tree was constructed with Phylip [64] using neighbor joining and the JTT exchange matrix as described in the text. The tree was drawn with TreeView [65].

Nucleotide Polymorphism in *V. destructor*

Life-history traits of *V. destructor* [5] that act to reduce genetic polymorphism within family lineages include male haploidy [25] and a predominance of full-sib mating. Genetic variation within a population can nonetheless be high in principle if populations are admixtures of distinct lineages. High-frequency polymorphisms, whether due to admixture or heterozygosity, can cause difficulties for shotgun assembly algorithms because they weaken the computational discrimination of allelic and non-allelic sequences. On the other hand, polymorphisms may be useful as genetic markers for population-genetic studies. It is therefore of interest to estimate levels of genetic polymorphism in the sequenced sample (~1,000 mites drawn from three adjacent colonies). We used the program SWAP454 [52] to estimate the occurrence of moderate- to high-frequency SNPs, i.e. those present in reads at a minimum ratio of 0.1 to the assembly reference base. SNP calls also required a minimum of two reads with the alternative base, but did not require reads in both directions. Polymorphisms meeting these parameters occurred at a rate of 6.2×10^{-5} per base pair. Given a median per-contig coverage of 5.0× in the final assembly (Table 1), our ability to detect low frequency polymorphisms is of course limited, but such polymorphisms contribute much less ambiguity to genome assembly.

To further investigate the potential for sequence polymorphism within a *V. destructor* population, we identified trinucleotide microsatellite loci in the genomic sequence and obtained amplification products for ten of these (see Methods). Consistent with our estimate of SNP frequency, we found no polymorphism at these loci in 65 individual mites collected from research apiaries at the USDA-ARS facility in Beltsville, Maryland, the source of the genome survey pool.

DISCUSSION

V. destructor is considered the most damaging honey-bee pest and has become widespread since its host shift from *A. cerana* less than a century ago. Resistance to common acaricides has already appeared [53], and the development of new control strategies are hampered by our limited knowledge of the *V. destructor - A. mellifera* interaction, particularly at the genetic level. The present genome survey makes available a large number of genic sequences for analysis and manipulation by the community of researchers. The contigs we obtained from low-coverage shotgun sequencing were short, as expected, such that few complete gene models are likely to be annotated from this data set. Nonetheless, our assembly enables the identification of genes of interest and the cloning of complete transcripts as they are needed. The assembly will also greatly assist the validation and annotation of transcriptome surveys and can support proteomic initiatives. We hope the resources provided here will aid investigators already tackling the problem of mite control with molecular methods as well as encourage the involvement of others.

Genetic approaches to the study of mite control are promising for a number of reasons. Successful reproduction of *Varroa* mites requires precise coordination with the care of honey bee brood and a sophisticated evasion of honey bee defences. Chemosensory genes are among those likely to play crucial roles in this process. At the phenotypic level, there is known variation among *Varroa*haplotype groups and species in their ability to successfully parasitize *A. mellifera* [16, 17], as well as known variation in the resistance of honey bee strains [14, 15, 54, 55]. While *Varroa* mites are not tractable to controlled crosses, associative mapping of traits such as reproductive success on different hosts remains an attractive possibility, as is the mapping of resistance traits in honey bee. Resequencing efforts are needed to uncover genetic variation that can be exploited for these purposes. Those efforts would also contribute to a better understanding of the demographic history of *V. destructor* and to species relationships within the genus. Polymorphic markers within haplotype groups would aid investigations of the population biology of the species, particularly in light of the difficulty of observing or manipulating *Varroa* mites in their habitat. For example, estimates of outcrossing and migration might be relevant to the design of new mitigation strategies, particularly if the evolution of resistance traits is expected. While consistent with previous work [23] that found low genetic polymorphism within the predominant *V. destructor* lineage, our analysis nonetheless enables genome-scale mining of markers for population-genetic studies.

Our analysis of conserved peptide blocks showed a higher rate of protein evolution within the*Varroa* lineage relative to *Ixodes*. It remains to be clarified whether this level of sequence divergence is characteristic of mites or peculiar to the *Varroa* lineage. Similarly, whether this result correlates with divergence in other aspects of genome evolution, such as exon structure and regulatory features, will be an important question to pursue as annotations improve for both species. As genomic resources for *V. destructor* continue to improve, opportunities for evolutionary comparisons with other arthropods will be enriched. Such comparisons are of tremendous value because they can expose conserved elements that might otherwise elude detection by direct experiment, and they reveal the relative rates at which various classes of homologous sequence diverge. *Varroa* mites and others in the Parasitiformes comprise a lineage that diverged from ixodid ticks over 300 million years ago [56], while the chelicerates as a whole branched from the insects and crustacea 725 million years ago [57]. Consequently, as a representative of the Parasitiformes, *Varroa* provides a key evolutionary landmark for comparative studies across arthropods currently targeted for genomic analyses.

An accessory goal of genome projects targeting arthropod pests is the identification of novel microbes and viruses that may be relevant to the epidemiology of vectored diseases, or that lend themselves to biocontrol programs. A significant finding of this study was the discovery of an actinomycete bacterium that infects *V. destructor* at intermediate frequencies (albeit presumably at high titer given its abundance in the genome sequence) but apparently does not infect *A. mellifera*at appreciable levels. However, these findings are preliminary and await a more systematic survey of infection among mites and bees. Future research should also be directed toward isolating this bacterium and assessing the fitness consequences of infection. Further characterization of the putative baculovirus identified in this survey is similarly a priority.

This work contributes to the relatively small body of genomic studies to date that have applied next-generation sequencing to a complex eukaryotic genome phylogenetically distant from other reference genomes. As the costs and technical requirements for genome sequencing continue to decline, such studies will undoubtedly become commonplace. In many cases, the sequencing of a single genotype or inbred group will not be feasible, and there may also be a significant metagenomic contribution from associated microbes. While these factors introduce new challenges, our results underscore the utility of these methods for rapidly advancing the study of non-model organisms.

CONCLUSIONS

Our results have provided general tools for the research community and novel directions for investigating the biology and control of *Varroa* mites. Ongoing development of *Varroa* genomic resources will be a boon for comparative genomics of under-represented arthropods, and will further enhance the honey bee and its associated pathogens as a model system for studying host-pathogen interactions.

METHODS

Genome-Size Estimation

Samples were prepared for flow cytometry as previously described [27]. For each replicate, the synganglion of a mature female *V.*

destructor was placed along with the head of a mature female *D. virilis* into a 2 ml tissue grinder (Kimble-Kontes) containing 1 ml of cold Galbraith buffer, and stroked 15 times with the A pestle to release nuclei. The preparation was filtered through 20 μm nylon and stained with prodidium iodide to a final concentration of 50 ppm. Stained samples were held on ice in the dark for 1-2 hr prior to analysis. The mean fluorescence of stained nuclei in replicate preparations of *Varroa* and *D. virilis* standard was quantified using a Coulter Epics Elite flow cytometer (Coulter Electronic), with excitation provided by a laser tuned at 488 nm and 25 mW. PI fluorescence at > 615 nm was detected by a photomultiplier screened by a long pass filter. To ensure that scoring included only intact nuclei free from cytoplasmic tags, counting was activated by red fluorescence (discrimination), and only (gated) nuclei with low forward and side scatter were included in the analysis. The positions of sample peaks relative to the *D. virilis* peak were verified by running samples without a standard. DNA content was determined from co-preparations as the ratio of the 2C *Varroa* peak to the 2C *D. virilis* peak times the 1C genome size of *D. virilis* (333 Mb, after [58]).

Sample Preparation and Sequencing

V. destructor mites were collected on two occasions for sequencing from infested colonies of the USDA-ARS Bee Research Laboratory apiaries in Beltsville, MD, USA in October, 2008. For each collection, 300 bees were placed into a 0.5 liter glass jar containing 40 g confectioner's sugar. The sugar and bees were thoroughly mixed by shaking the jar for 30 s. The sugar and phoretic mites were then separated from their bee hosts by shaking the jar contents through a 1 mm wire mesh placed at the jar opening. Mites were shaken into a small water bath, which was then poured through a cheesecloth filter and rinsed twice with sterile water to remove residual sugar. Live mites were then picked onto sterile tissue paper and frozen at -80°C until nucleic acid extraction.

To obtain sufficient high-quality DNA for six pyrosequencing runs, three separate DNA extractions were made from the collected mites. DNA from one sample of ~400 mites was extracted with DNAzol (Invitrogen) following the manufacturer's instructions. A second sample of ~400 mites was homogenized in 800 μl proteinase K buffer (10 mM NaCl, 10 mM Tris, 50 mM ethylenediamenetetracetic acid (EDTA), and

10 μg/ul proteinase K) and incubated 60 min at 55°C, vortexing every 20 min. Afterwards, 180 μl of 8 M potassium acetate was added and the sample incubated on ice for 30 min. After high-speed centrifugation, DNA was precipitated from the supernatant with ethanol and re-suspended in distilled, deionized water. A third sample of ~200 mites was homogenized in 500 μl hexadecyltrimethylammonuim bromide (CTAB) buffer (100 mM Tris-HCl at pH 8.0, 20 mM EDTA, 1.4 M NaCl, 2% CTAB, and 0.2% -mercaptoethanol) and incubated 60 min at 65°C, vortexing every 20 min. An equal volume of 24:1 chloroform:isoamyl alcohol was then added and the sample centrifuged at high speed. DNA was precipitated from the supernatant with isopropanol and re-suspended in distilled, deionized water. For all extractions, DNA concentration and quality were evaluated with a Nanodrop ND-8000 spectrophotometer and were found to be comparable. DNA quality was also checked by gel electrophoresis.

Pyrosequencing was performed at the Institute for Genome Sciences, University of Maryland School of Medicine, on a Genome Sequencer FLX instrument (454 LifeSciences) using GS-FLX titanium reagents. DNA was prepared for emulsion PCR according to the manufacturer›s protocols.

Assembly and Analysis

Contigs were assembled with the CABOG package of Celera Assembler version 5.2 [31]. The sequences were assembled iteratively, adjusting the assumed error rate incrementally between 0% and 6%. The assembly selected for analysis used a 1.5% error rate because this value maximized the length of the longest contig (18.7 Kbp). The weighted median N50 contig size was relatively stable across iterations at ~2.1 Kbp, meaning that half of the assembled bases were consistently contained in contigs of this size or larger across the different assemblies.

Contigs were screened to identify sequences of organisms that were considered potential contaminants *a priori*. For example, a number of contigs were found to be nearly identical to the honey bee reference genome [59]. These fragments included low-copy genic sequences as well as ribosomal, nongenic, and mitochondrial sequence. PCR primers specific to *A. mellifera* sequences amplified genomic DNA extracted from adult female mites but not from embryos (data not shown), implying that the source of *A. mellifera* contamination is bee hemolymph consumed by mites. In contrast, searches against the genome sequences [21] of chalkbrood (*Ascosphaera apis*), a fungal pathogen of honey bees, and American foulbrood (*Paenibacillus larvae*), a bacterial pathogen, did not reveal the presence of these spore-dispersed microbes in the sample.

All contigs were also screened for general bacterial contamination by searching ORFs of 90 codons or more against the RAST [60] seed database of bacterial sequences. ORFs were identified with the *getorf* program of the EMBOSS package [61]. ORFs with significant matches to the RAST database were then searched by BLASTP against the full GenBank nr database to determine the most closely matching organism. Candidate microbial sequences are summarized in Table 6. As expected (see Results), matches to the phylum Actinobacteria are the vast majority. Among the small number of BLASTP matches to other bacterial groups, the genera *Burkholderia* (-proteobacteria) and *Pseudomonas* (-proteobacteria) were the most represented taxa.

Table 6: Distribution of contigs that were designated bacterial by BLAST analysis, sorted by phylum

Phylum	Number of contigs
Actinomycete	1035
-proteobacteria	10
Aquificae	1
Bacteroides	2
-proteobacteria	13
Cyanobacteria	3
-proteobacteria	3
Euryarachaeota	1
Firmicutes	4
-proteobacteria	12

Cornman et al. BMC Genomics 2010 11:602, doi: 10.1186/1471-2164-11-602

Because many organisms show distinct patterns of codon usage [62], we compared codon usage for ORFs from unfiltered contigs (putatively *V. destructor* with those from putatively bacterial contigs.

Only ORFs with BLAST-supported homology to GenBank sequences were used for this comparison. We used the program INCA [63] to plot the codon-usage statistic 'B' of [62] as a function of third-position G+C (GC3) content The value of B for a given ORF is a measure of how similar its codon usage is to the overall codon usage in the data set. GC3 is considered here because third positions are much less constrained by protein function than first and second positions, and thus more indicative of background composition biases. The distinct patterns observed for the two groups of ORFs provide complementary evidence that these sequences do in fact derive from different organisms. The plot also shows that a few ORFs from contigs considered to be *Varroa* by our filtering methods may in fact be bacterial in nature and merit further evaluation. Of course, BLAST-supported ORFs are only proxies for transcripts and thus individual points may be highly inaccurate. In general, however, genic sequences that are putatively from *V. destructor*have a cohesive pattern of codon usage that can be distinguished form at least some bacterial contaminants, regardless of detectable homology.

To confirm the presence in mites of the actinomycete bacterium and DNA virus identified in the assembly, individual eggs, nymphs, female adults, and male adults were collected from parasitized honey bee pupae. DNA for PCR was extracted from individual samples by grinding them in 200 μl of 5% Chelex-100 solution (Bio-Rad), incubating at 65°C for 30 min, pelleting the mixture by high-speed centrifugation, and taking a 1:10 dilution in water of the resulting supernatant. Primers for the putative actinomycete TIF3 locus were CCGATCTCGACCTTGTGGAA (forward) and CTCGGAACATGATCGTCACC (reverse), and for the ABC locus were GAGGTCCTCGTCTCCGAATG (forward) and CGATGTCCTGGATCCTCTGG (reverse). The amplified TIF3 product was confirmed by Sanger sequencing (GenBank:GU365869). Primers designed to amplify the putative Baculovirus targeted a ribonucleotide reductase small subunit gene (forward ACGAACGACTATCTAGCCATGAAC and reverse GTCCGTTTCGGAGTGCATGAC) and a thymidylate synthase gene (forward CGCATGTACCAACAACTCGTAC and reverse CACAGTTGGTGTAGCGCAGT). The identities of these products were also confirmed by Sanger sequencing (GenBank:GU980896 and GenBank:GU980897, respectively). All PCR reactions were performed using standard reagents and thermocycler protocols, with an annealing temperature of 54°C.

To identify conserved peptide blocks, we first identified *V. destructor* ORFs that were reciprocal best BLASTP matches with *I. scapularis* predicted peptides. These were then used to identify the closest homologs in *Da. pulex, Dr. melanogaster,* and *P. humanus*. Sequences were aligned by ClustalW and then trimmed to include only blocks of well conserved, unambiguously aligned sequence for which we could have reasonable confidence of orthology. Genetic-distance matrices were calculated for each block with the *protdist* program of the PHYLIP package [64], weighted by alignment length and summed, then normalized to a maximum distance of one. Unrooted dendrograms were constructed with the fitch and neighbor programs of the PHYLIP package, giving virtually identical branch lengths; the neighbor-joining dendrogram is shown in Figure 5.

Microsatellite loci used to assess polymorphism levels in *V. destructor* are characterized in. PCR amplifications consisted of 1 U Taq DNA polymerase with appropriate buffer, 1 mM dNTP, 2 mM MgCl2, 0.2 μM of each forward and reverse primer in a final reaction volume of 5 μl. Fluorescently labeled primers were mixed with unlabeled primers at a 12:20 ratio. Thermocycling was performed as follows: 96°C for 2 min., then 3 cycles of 96°C for 30 sec., 60°C for 30 sec. (-1°C/Cycle), 65°C for 1 min., followed by 35 cycles of 96°C for 30 sec., 56°C for 30 sec., 65°C for 1 min, and a final extension at 65°C for 2 min. PCR products were diluted 1:20 and 1 μl of this dilution was added to 10 μL formamide containing the LIZ size standard. Products were analyzed by capillary electrophoresis using an Applied Biosystems 3730XL instrument. Allele sizes were scored using ABI GeneMapper version 3.7 (Applied Biosystems).

AUTHORS' CONTRIBUTIONS

RSC led the bioinformatic analyses and writing. JDE designed and coordinated the project and helped with analyses and writing. MCS generated the genome assembly. JSS performed the genome size estimation. YPC and JSP participated in project design and the collection of biological materials. GH, LB, CE, DA, and CMG aided in data analysis and manuscript preparation. All authors read and approved the final manuscript.

ACKNOWLEDGEMENTS

Financial support for this survey and an ongoing full genome sequencing project provided by USDA-ARS and the USDA National Institute of Food and Agriculture (grant 2009-05254 to JDE, GH, CE and LB). Additional support was provided by NIH grant R01-LM006845 and NSF grant NSF IIS-084494 to the University of Maryland (Steven Salzberg) and the University of Florida Department of Agriculture (JDE). We thank Antoinette Betschart and Kevin Hackett, USDA-ARS, for logistical support, Dawn Lopez for technical support, and the Institute for Genomic Sciences, University of Maryland, Baltimore, for sequencing. The helpful critique of three anonymous reviewers greatly improved the manuscript.

REFERENCES

1. Morse RA, Calderone NW: The value of honey bees as pollinators of U.S. crops in 2000. *Bee Culture* 2000, 128:1-15.
2. Gallai N, Salles JM, Settele J, Vaissiere BE: Economic valuation of the vulnerability of world agriculture confronted with pollinator decline. *Ecological Economics* 2009, 68:810-821
3. Aizen MA, Harder LD: The global stock of domesticated honey bees is growing slower than agricultural demand for pollination. *Current Biology* 2009, 19:915-918.
4. Cox-Foster DL, Conlan S, Holmes EC, Palacios G, Evans JD, Moran NA, Quan PL, Briese T, Hornig M, Geiser DM, *et al.*: A metagenomic survey of microbes in honey bee colony collapse disorder. *Science* 2007, 318:283-287
5. Rosenkranz P, Aumeier P, Ziegelmann B: Biology and control of *Varroa destructorJ Invertebr Pathol* 103(Suppl 1):S96-119
6. Anderson DL, Trueman JWH: *Varroa jacobsoni* (Acari: Varroidae) is more than one species. *Experimental and Applied Acarology* 2000, 24:165-189
7. Johnson RM, Evans JD, Robinson GE, Berenbaum MR: Changes in transcript abundance relating to colony collapse disorder in honey bees (*Apis mellifera*). *Proc Natl Acad Sci USA* 2009, 106:14790-14795.

8. Highfield AC, El Nagar A, Mackinder LC, Noel LM, Hall MJ, Martin SJ, Schroeder DC: Deformed wing virus implicated in overwintering honeybee colony losses. *Appl Environ Microbiol* 2009, 75:7212-7220.
9. Yang X, Cox-Foster D: Effects of parasitization by *Varroa destructor* on survivorship and physiological traits of *Apis mellifera* in correlation with viral incidence and microbial challenge. *Parasitology* 2007, 134:405-412.
10. Boecking O, Genersch E: Varroosis - The ongoing crisis in bee keeping. *Journal fur Verbraucherschutz und Lebensmittelsicherheit* 2008, 3:221-228.
11. Sammataro D, Gerson U, Needham G: Parasitic mites of honey bees: life history, implications, and impact. *Annu Rev Entomol* 2000, 45:519-548
12. Kuenen LPS, Calderone NW: Transfers of *Varroa* mites from newly emerged bees: Preferences for age- and function-specific adult bees (Hymenoptera: Apidae). *Journal of Insect Behavior* 1997, 10:213-228.
13. Kraus B: Factors influencing host choice of the honey bee parasite *Varroa jacobsoni*Oud. *Experimental and Applied Acarology* 1994, 18:435-443.
14. De Guzman LI, Rinderer TE, Frake AM: Comparative reproduction of *Varroa destructor*in different types of Russian and Italian honey bee combs. *Experimental and Applied Acarology* 2008, 44:227-238.
15. Rinderer TE, De Guzman LI, Delatte GT, Harper C: An evaluation of ARS Russian honey bees in combination with other methods for the control of *Varroa* mites. *American Bee Journal* 2003, 143:410-413.
16. Anderson DL: Non-reproduction of *Varroa jacobsoni* in *Apis mellifera* colonies in Papua New Guinea and Indonesia. *Apidologie* 1994, 25:412-421.
17. Anderson DL, Fuchs S: Two genetically distinct populations of *Varroa jacobsoni* with contrasting reproductive abilities on *Apis mellifera*. *Journal of Apicultural Research* 1998, 37:69-78.
18. Navajas M, Anderson DL, De Guzman LI, Huang ZY, Clement J, Zhou T, Le Conte Y: New Asian types of *Varroa destructor* a potential new threat for world apiculture. *Apidologie* 2010.

19. NCBI Genome Project ID 33467 [http://www.ncbi.nlm.nih.gov/genomeprj?term=Varroa%20destructor]
20. Cornman RS, Chen YP, Schatz MC, Street C, Zhao Y, Desany B, Egholm M, Hutchison S, Pettis JS, Lipkin WI, *et al.*: Genomic analyses of the microsporidian *Nosema ceranae*, an emergent pathogen of honey bees. *PLoS Pathog* 2009, 5:e1000466.
21. Qin X, Evans JD, Aronstein KA, Murray KD, Weinstock GM: Genome sequences of the honey bee pathogens *Paenibacillus larvae* and *Ascosphaera apis*. *Insect Molecular Biology* 2006, 15:715-718.
22. De Guzman LI, Rinderer TE, Stelzer JA: Occurrence of two genotypes of *Varroa jacobsoni* oud. in North America. *Apidologie* 1999, 30:31-36.
23. Solignac M, Cornuet J-M, Vautrin D, Le Conte Y, Anderson D, Evans J, Cros-Arteil S, Navajas M: The invasive Korea and Japan types of *Varroa destructor*, ectoparasitic mites of the Western honeybee (*Apis mellifera*), are two partly isolated clones. *Proceedings of the Royal Society - Biological Sciences (Series B)* 2005, 272:411-419.
24. Anderson DL: Variation in the parasitic bee mite *Varroa jacobsoni*Oud.*Apidologie* 2000, 31:281-292.
25. Steiner J, Grasas Pompolo S, Takahashi CS, Gonsalves LS: Cytogenetics of the acarid*Varroa jacobsoni*. *Brazilian Journal of Genetics* 1982, 4:841-844.
26. Pagel Van Zee J, Geraci NS, Guerrero FD, Wikel SK, Stuart JJ, Nene VM, Hill CA: Tick genomics: the *Ixodes* genome project and beyond. *Int J Parasitol* 2007, 37:1297-1305.
27. Geraci NS, Johnston JS, Robinson JP, Wikel SK, Hill CA: Variation in genome size of argasid and ixodid ticks. *Insect Biochemistry and Molecular Biology* 2007, 37:399-408.
28. Grbic M, Khila A, Lee KZ, Bjelica A, Grbic V, Whistlecraft J, Verdon L, Navajas M, Nagy L: Mity model: *Tetranychus urticae*, a candidate for chelicerate model organism *Bioessays* 2007, 29:489-496.
29. wFleaBase [http://wfleabase.org/]
30. The International Aphid Genomics Consortium: Genome sequence of the pea aphid*Acyrthosiphon pisum*. *PLoS Biol* 2010, 8:e1000313.

31. Miller JR, Delcher AL, Koren S, Venter E, Walenz BP, Brownley A, Johnson J, Li K, Mobarry C, Sutton G: Aggressive assembly of pyrosequencing reads with mates. *Bioinformatics* 2008, 24:2818-2824
32. Lander ES, Waterman MS: Genomic mapping by fingerprinting random clones: a mathematical analysis.*Genomics* 1988, 2:231-329.
33. Tyson GW, Chapman J, Hugenholtz P, Allen EE, Ram RJ, Richardson PM, Solovyev VV, Rubin EM, Rokhsar DS, Banfield JF: Community structure and metabolism through reconstruction of microbial genomes from the environment.*Nature* 2004, 428:37-43.
34. Gotz S, Garcia-Gomez JM, Terol J, Williams TD, Nagaraj SH, Nueda MJ, Robles M, Talon M, Dopazo J, Conesa A: High-throughput functional annotation and data mining with the Blast2GO suite. *Nucleic Acids Res* 2008, 36:3420-3435.
35. Klepzig KD, Adams AS, Handelsman J, Raffa KF: Symbioses: a key driver of insect physiological processes, ecological interactions, evolutionary diversification, and impacts on humans. *Environ Entomol* 2009, 38:67-77.
36. Finn RD, Tate J, Mistry J, Coggill PC, Sammut SJ, Hotz HR, Ceric G, Forslund K, Eddy SR, Sonnhammer EL, *et al.*: The Pfam protein families database. *Nucleic Acids Res* 2008, 36:D281-288.
37. Baculovirus Molecular Biology [http://www.ncbi.nlm.nih.gov/bookshelf/br.fcgi?book=bacvir]
38. Rosell G, Quero C, Coll J, Guerrero A: Biorational insecticides in pest management. *Journal of Pesticide Science* 2008, 33:103-121
39. Herniou EA, Olszewski JA, Cory JS, O'Reilly DR: The genome sequence and evolution of baculoviruses. *Annu Rev Entomol* 2003, 48:211-234.
40. Ogata H, Goto S, Sato K, Fujibuchi W, Bono H, Kanehisa M: KEGG: Kyoto Encyclopedia of Genes and Genomes. *Nucleic Acids Res* 1999, 27:29-34.
41. Vectorbase [http://www.vectorbase.org]
42. Parra G, Bradnam K, Ning Z, Keane T, Korf I: Assessing the gene space in draft genomes. *Nucleic Acids Res* 2009, 37:289-297.

43. HMMER3: a new generation of sequence homology search software[http://hmmer.janelia.org]
44. RepeatMasker Open-3.0 [http://www.repeatmasker.org]
45. Volff JN: Turning junk into gold: domestication of transposable elements and the creation of new genes in eukaryotes. *Bioessays* 2006, 28:913-922.
46. Atkinson PW, Pinkerton AC, O'Brochta DA: Genetic transformation systems in insects. *Annu Rev Entomol* 2001, 46:317-346.
47. Jeyaprakash A, Hoy MA: First divergence time estimate of spiders, scorpions, mites and ticks (subphylum: Chelicerata) inferred from mitochondrial phylogeny. *Exp Appl Acarol* 2009, 47:1-18.
48. Jones DT, Taylor WR, Thornton JM: The rapid generation of mutation data matrices from protein sequences. *Computer Applications in the Biosciences* 1992, 8:275-282.
49. Felsenstein J: *Inferring phylogenies.* Sunderland, Mass.: Sinauer; 2004.
50. Regier JC, Shultz JW, Zwick A, Hussey A, Ball B, Wetzer R, Martin JW, Cunningham CW:Arthropod relationships revealed by phylogenomic analysis of nuclear protein-coding sequences. *Nature* 463:1079-1083.
51. Wheeler WC, Whiting M, Wheeler QD, Carpenter JM: The phylogeny of the extant hexapod orders. *Cladistics* 2005, 17:113-169.
52. Brockman W, Alvarez P, Young S, Garber M, Giannoukos G, Lee WL, Russ C, Lander ES, Nusbaum C, Jaffe DB: Quality scores and SNP detection in sequencing-by-synthesis systems. *Genome Res* 2008, 18:763-770.
53. Pettis JS: A scientific note on *Varroa destructor*resistance to coumaphos in the United States. *Apidologie* 2004, 35:91-92.
54. Guzman-Novoa E, Vandame R, Arechavaleta ME: Susceptibility of European and Africanized honey bees (*Apis mellifera* L.) to *Varroa jacobsoni* Oud. in Mexico. *Apidologie* 1999, 30:173-182.
55. Moretto G, Mello LJd Jr: *Varroa jacobsoni* infestation of adult Africanized and Italian honey bees (*Apis mellifera*) in mixed colonies in Brazil. *Genetics and Molecular Biology* 1999, 22:321-323.

56. Jeyaprakash A, Hoy MA: First divergence time estimate of spiders, scorpions, mites and ticks (subphylum: Chelicerata) inferred from mitochondrial phylogeny. *Experimental and Applied Acarology* 2009, 47:1-18.
57. Pisani D, Poling LL, Lyons-Weiler M, Hedges SB: The colonization of land by animals: molecular phylogeny and divergence times among arthropods. *BMC Biology* 2004, 2:1.
58. Gregory TR, Johnston JS: Genome size diversity in the family Drosophilidae. *Heredity* 2008, 101:228-238.
59. Honey Bee Genome Sequencing Consortium: Insights into social insects from the genome of the honeybee *Apis mellifera*. *Nature* 2006, 443:931-949.
60. Aziz RK, Bartels D, Best AA, DeJongh M, Disz T, Edwards RA, Formsma K, Gerdes S, Glass EM, Kubal M, *et al.*: The RAST Server: rapid annotations using subsystems technology. *BMC Genomics* 2008, 9:75.
61. Rice P, Longden I, Bleasby A: EMBOSS: the European Molecular Biology Open Software Suite. *Trends Genet* 2000, 16:276-277.
62. Karlin S, Mrazek J, Campbell AM: Codon usages in different gene classes of the*Escherichia coli* genome. *Mol Microbiol* 1998, 29:1341-1355.
63. Supek F, Vlahovicek K: INCA: synonymous codon usage analysis and clustering by means of self-organizing map. *Bioinformatics* 2004, 20:2329-2330.
64. Felsenstein J: PHYLIP - Phylogeny Inference Package (Version 3.2). *Cladistics* 1989, 5:164-166.
65. Page RD: TreeView: an application to display phylogenetic trees on personal computers. *Comput Appl Biosci* 1996, 12:357-358.
66. Interpro [http://www.ebi.ac.uk/interpro/]
67. Uniprot [http://www.uniprot.org]
68. Jurka J, Kapitonov VV, Pavlicek A, Klonowski P, Kohany O, Walichiewicz J: Repbase Update, a database of eukaryotic repetitive elements. *Cytogenet Genome Res* 2005, 110:462-467.

Chapter 9

Repellent Activities of Essential Oils of Some Plants Used Traditionally to Control the Brown Ear Tick, Rhipicephalus Appendiculatus

Wycliffe Wanzala[1,2], Ahmed Hassanali[2,3], Wolfgang Richard Mukabana[2,4], and Willem Takken[5]

[1]Department of Biological Sciences, School of Pure and Applied Sciences, South Eastern Kenya University, P.O. Box 170-90200, Kitui, Kenya

[2]Behavioural and Chemical Ecology Department, International Centre of Insect Physiology and Ecology, African Insect Science for Food and Health, P.O. Box 30772-00100-GPO, Nairobi, Kenya

[3]Chemistry Department, School of Pure and Applied Sciences, Kenyatta University, P.O. Box 43844-00100-GPO, Nairobi, Kenya

[4]School of Biological Sciences, University of Nairobi, P.O. Box 30197-00100-GPO, Nairobi, Kenya

[5]Laboratory of Entomology, Wageningen University and Research Centre, P.O. Box 8031, 6700 EH Wageningen, The Netherlands

ABSTRACT

Essential oils of eight plants, selected after an ethnobotanical survey conducted in Bukusu community in Bungoma County, western Kenya (Tagetes minuta, Tithonia diversifolia, Juniperus procera, Solanecio mannii, Senna didymobotrya, Lantana camara, Securidaca longepedunculata, and Hoslundia opposita), were initially screened (at two doses) for their repellence against brown ear tick, Rhipicephalus appendiculatus, using a dual-choice climbing assay. The oils of T. minuta and T. diversifolia were then selected for more detailed study. Dose-response evaluations of these oils showed that T. minuta oil was more repellent (RD_{50} = 0.0021 mg) than that of T. diversifolia (RD_{50} = 0.263 mg). Gas chromatography-linked mass spectrometric (GC-MS) analyses showed different compositions of the two oils. T. minuta oil is comprised mainly of cis-ocimene (43.78%), dihydrotagetone (16.71%), piperitenone (10.15%), trans-tagetone (8.67%), 3,9-epoxy-p-mentha-1,8(10)diene (6.47%), -ocimene (3.25%), and cis-tagetone (1.95%), whereas T. diversifolia oil is comprised mainly of -pinene (63.64%), -pinene (15.00%), isocaryophyllene (7.62%), nerolidol (3.70%), 1-tridecanol (1.75%), limonene (1.52%), and sabinene (1.00%). The results provide scientific rationale for traditional use of raw products of these plants in controlling livestock ticks by the Bukusu community and lay down some groundwork for exploiting partially refined products such as essential oils of these plants in protecting cattle against infestations with R. appendiculatus.

INTRODUCTION

In sub-Saharan Africa, East Coast fever (ECF), caused by Theileria parva parva, Theiler, 1904, and transmitted by the brown ear tick, Rhipicephalus appendiculatus, Neumann, 1901, is one of the major constraints to the development of the livestock industry [1, 2]. Of the estimated 12.7 million head of cattle (both indigenous and exotic), 76% are at risk to ECF [3]. The disease is associated with up to 10%

mortality in zebu calves in ECF endemic areas and can cause up to 100% mortality in susceptible exotic and indigenous breeds [3, 4].

Prevention, control, and management of both vector and pathogen have continued to rely heavily on the application of synthetic chemical acaricides on the host since their introduction in 1902 in sub-Saharan Africa [5]. However, this has proved to be costly and unsustainable in a number of ways [1]. The acaricides can eliminate ticks from the host, but they do not prevent continued reinfestation from the source environment, where ticks spend 90% of their life. For effective management of harmful ticks, an integrated combination of tactics may need to be put in place that controls ticks on individual hosts as well as in the host environment in order to prevent host reinfestation during grazing. One possible strategy would be to use tick repellents on the host and tick-repellent plants in the pasture (host environment), combined with plants that are attractive to ticks such as Acalypha fruticosa Forssk. var. villosa Hutch (Family: Euphorbiaceae) surrounding the pasture land so as to develop a "push-pull" tick manipulation system [6, 7]. Although the proposed strategy appears complex, it may be possible to achieve in zero/semizero grazing, small-scale free-range, and tethering livestock farming systems. In others, such as pastoralism and large-scale livestock farming systems, the deployment of well-formulated repellents dispensed from controlled-release dispensers may be more practical.

N,N-diethyl-3-methylbenzamide (DEET) is still considered the best available product, repelling a wide variety of insects, ticks, and mites [8]. Though DEET is not expected to bioaccumulate, the amounts present in the environment have been shown to be toxic to some species of zooplankton and fish [9, 10]. In humans, the repellent may cause insomnia, mood disturbances, impaired cognitive functions, seizures, toxic encephalopathy, and allergic reactions [11–13]. This has led to a search for alternative eco-friendly and effective repellents.

The potential of some local plants and plant products to repel ticks from grazing areas and host animals, respectively, has been demonstrated previously [14–18]. Melinis minutiflora (molasses grass), a tropical grass already in use as livestock fodder [19, 20], covers crop and mulch [21] and for thatching houses [22], it has been shown to be toxic [15] and repellent to ticks [14, 23, 24] as well as insects and snakes [21]. One study demonstrating potential of molasses grass to control Amblyomma variegatum and R. appendiculatus, vectors of the livestock diseases heartwater (cowdriosis) and ECF, respectively, has been reported [24]. In addition, several other Kenyan local shrubs, including Cleome hirta and Gynandropsis gynandra, have demonstrated potential as tick-repellent pasture plants [25–27].

A number of studies have shown that plant-based repellents can be comparable to DEET or even better [28–32]. One commercial repellent product is the Flea and Tick Granular Repellent, which is made from essential oils of cedar, cinnamon, mint, and lemon grass; it has a pleasant odour and can be safely used outdoors for flea and tick control [33]. Essential oils of a number of other plants have been shown to be repellent to ticks. These include Commiphora erythraea and C. myrrh [34], Cleome monophylla [35], Ocimum suave [36], Cleome hirta [27], and G. gynandra [37].

Use of tick-repellent plants in pasture lands or essential oils on hosts and their integration with other off-host or on-host tick control measures could be practical and provide economic ways of controlling not only livestock ticks but also arthropod vectors [38–40]. In our previous survey of livestock tick control ethnopractices among Bukusu community in Bungoma district, western Kenya, we found widespread use of ethnobotanicals derived from local/native plants to control tick infestations on cattle [41]. Blends of botanicals from one or more plants are used either as on-host suspensions or burnt and smoke used to fumigate cattle. Our follow up objective has been to assess the repellence of essential oils of some of these plants against R. appendiculatus adults in the laboratory, to characterize the chemical constituent profiles of the more repellent ones, and then to initiate both off- and on-host evaluation of their efficacy in controlling the ticks in the field. In the present paper, we report the results obtained from repellence assays of essential oils of 8 plants against R. appendiculatus adults and results of a more detailed study of two selected plants, Tagetes minuta L. and Tithonia diversifolia (Hemsl.) A. Gray.

MATERIALS AND METHODS

Selection of Eight Plant Species

An ethnobotanical survey was previously conducted in the Bukusu community in Bungoma County, western Kenya, along the southern slopes and foothills of Mount Elgon at altitudes ranging from about 1,300 m in the south to about 3,500 m in the north [41]. The County is located between latitude 0°25'S and 0°53'N and longitude 34°21'W and 35°04'E. Specimens of ~157 plant species, which were documented to have varied effects on livestock ticks [41], were collected for taxonomic examination at the herbarium of the School of Biological Sciences, University of Nairobi, Kenya. The potential efficacy of each plant species in protecting cattle against tick infestations was assessed following a four-level protocol proposed by Heinrich and coworkers [42], and eight plant species were selected for initial laboratory screening [41, 43]. Voucher specimens of these plants were deposited at the University of Nairobi Herbarium, and comprised of Tagetes minuta L. (029-BGM-Mwi/2002), Tithonia diversifolia (Hemsl.) A. Gray (015-BGM-Muf/2002), Juniperus procera Endl. (134-BGM-Elg/2002), Solanecio manii (Hook. f.) C. Jeffrey. (106-BGM-Mwi/2002), Senna didymobotrya (Fresen.) H. S. Irwin and Barneby (132-BGM-Web/2002), Lantana camara L. (043-BGM-Mwi/2002), Securidaca longepedunculata Fres. (018-BGM-Mec/2002), and Hoslundia opposita Vahl. (133-BGM-Bul/2002).

Experimental Ticks

The ticks used (the brown ear tick, Rhipicephalus appendiculatus Neumann, 1901) were obtained from the colonies at the International Livestock Research Institute (ILRI) and bred in the insectary at ICIPE, Nairobi, Kenya. Rearing conditions and management of ticks were as described previously [44, 45]. All the experiments were conducted using the newly emerged adult ticks of mixed sexes.

Isolation of Essential Oils

The aerial parts of each of the eight plants were collected from the southern slopes and foothills of Mount Elgon in western Kenya during the month of August and allowed to dry in a well-ventilated room for 1-2 weeks. Each plant material was cut into small pieces and about 1 kg was hydrodistilled using a Clevenger-type apparatus for 8 h [46]. Essential oil of each plant was collected in 2 mL vials and stored at −20°C in a freezer until required for bioassays or analyses.

Dual-Choice Repellence Assays

A dual-choice tick repellence climbing assay [47] that exploits the behaviour of R. appendiculatus to climb up grass stems to await potential hosts passing by [48, 49] was used. The repellence of essential oils of the eight plants against R. appendiculatus was first compared at 0.1 mg and 50 mg doses. The most repellent oils (that ofT. minuta and that of T. diversifolia) were then selected for more detailed study. These plants are also highly ranked by livestock holders of Bukusu community in livestock tick prevention and control [41]. The oils of the plants were diluted serially with dichloromethane (GC grade) to provide 0.5 mg to 0.00005 mg/10 µL of solutions. An aliquot of 20 µL of each dose was applied to filter paper strip on the glass tubes, with an equivalent volume of dichloromethane added to the control filter paper strip. The-set up was allowed to equilibrate for 30 min before five adult R. appendiculatus of mixed age and sex were released on the base of the climbing set-up assay [47]. Observations were made over a 1-hour period, and the number of ticks above the filter paper strip on the control glass tube (Nc) and on the glass tube with test materials (Nt) was recorded at 15, 30, 45, and 60 min. Twenty replicates for each dose were carried out, each time with fresh, naïve adult ticks. Initial comparison of the responses of ticks in the set-up with and without residual dichloromethane on both sides showed no bias for either side and no effects of the residual solvent on the adult ticks. The repellency of each dose was calculated using the formula: (number of ticks in control arm − number of ticks in treated arm/total responding ticks) × 100. Dose-response data were subjected to probit analysis using the % repellencies from the replicated experiments [47].

Determination of the Composition of t. Minuta and t. Diversifolia Essential Oils

GC-MS analyses of the two oils were performed with a VG Masslab 12-250 quadruple gas chromatography-mass spectrometer. Chromatographic separations were achieved using a fused silica capillary column (Hewlett Packard, 50 m × 0.32 mm ID) coated with Carbowax 20 M (0.3 µm film thickness) with helium as the carrier gas. All the GC-MS analyses were made in the splitless mode with helium as the carrier gas. The oven temperature was programmed from 60°C for 7 min, to 120°C at 5°C per min, then to 180°C at 10°C per min, and finally to 220°C at 20°C per min, where it was maintained for l0 min. Constituents of the essential oils were identified by analysis of their mass spectra, direct comparison of these with those in the Wiley NBS and NIST databases, and coinjections with authentic standards (from Sigma Chemical Company, Poole, UK and Aldrich Chemical Company, Gillingham, UK) on a Hewlett Packard HP 5890A Gas Chromatograph equipped with a flame ionization detector (at 230°C). A fused silica capillary column (Hewlett Packard, 50 m × 0.22 mm × 0.33 mm CD) coated with methyl silicon (0.3 µm film thickness) was used with nitrogen as the carrier gas. All GC analyses were performed in the splitless mode with the injector temperature at 270°C and oven temperature programme similar to that in GC-MS analyses.

Data Analysis

Dose-response data were subjected to simple regression and probit analysis using the percent repellency values obtained from replicated experiments and a regression model developed based on

$$\text{Probit}[\Pi(\text{dose1})] = \beta_0 + \beta_1 x + \in, \quad (1)$$

where β_0 is the coefficient of the model representing y- intercept, β_1 is the coefficient of themodel representing dose1, x is the various concentrations of essential oils, dose1 is the Log10 (dose), $\in$ is the error term (residual term) representing the difference between the actual observed value and that predicted by the model (the predictor (regressor) variable, x is the dose of the essential oil), and is the repellency probability.

Student-Newman-Keuls H test was used to compare the mean values of repellency obtained for various doses of the repellent effects [50]. Percent repellency values were transformed into probabilities, while essential oil doses were transformed into logarithms to base 10 and lines for regression models fitted using R software for Microsoft windows. These models were used to estimate repellent effects of the two essential oils at RD_{50} and RD_{75} [8, 51].

RESULTS

Screening of the Essential Oils Isolated from the Selected Eight Plant Species

The results of repellency tests following the screening of the essential oils isolated from the eight plants (T. minuta, T. diversifolia, J. procera, S. mannii, S. didymobotrya, L. camara, S. longepedunculata, and H. opposita) at 0.1 mg and 50 mg doses are shown in Figure 1. Some variation was found in the repellent effect of the essential oils at the two doses with that of S. longepedunculata showing the least repellent effect at both doses and that of T. minuta showing the highest repellent effect at the lower dose (80.1 ± 4.7%). The essential oil ofT. minuta and one of the other six plants (T. diversifolia) were therefore selected for more detailed bioassay.

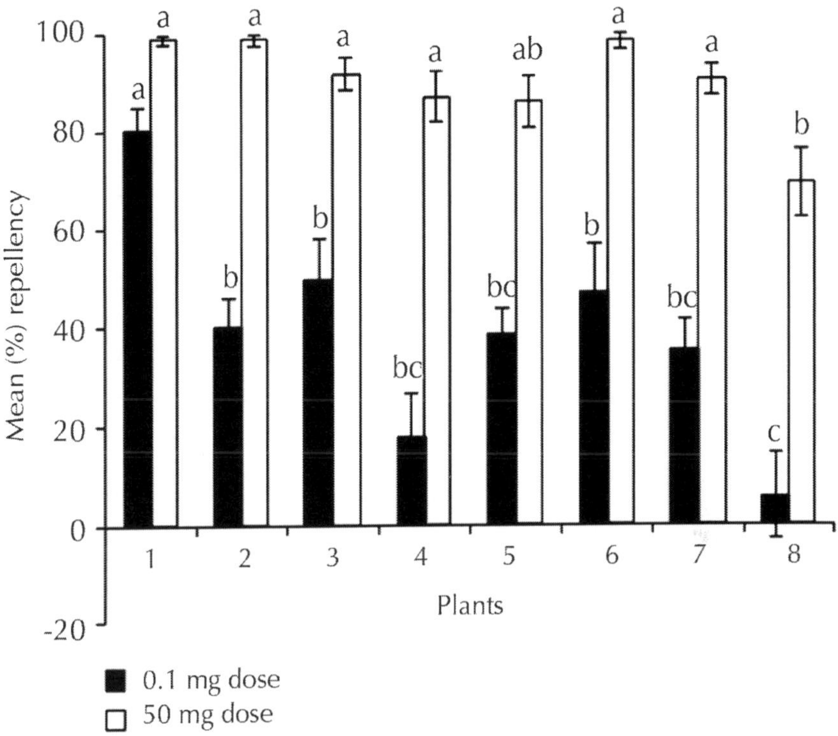

Figure 1: The repellent effect of essential oils of eight plants at doses of 0.1 mg and 50 mg (neat oil) against newly emerged Rhipicephalus appendiculatus adults. Plant species 1 isTagetes minuta, 2 is Tithonia diversifolia, 3 is Hoslundia opposita, 4 is Solanecio mannii, 5 is Lantana camara, 6 is Juniperus procera, 7 is Senna didymobotrya and 8 is Securidaca longepedunculata. For a given repellent dose, means capped by the same alphabetical letters are not significantly different at P<0.0001 (Student-Newman-Keuls H test).

Dose-Response Repellency of the Essential Oils of T. Minuta and T. Diversifolia

The repellence of the two essential oils at different doses is shown in Figures 2(a) and 2(b). The essential oil ofT. minuta was found to be significantly more repellent than that of T. diversifolia at all corresponding doses (). In both the essential oils of T. minuta and T. diversifolia, there was significant correlation between repellence and dose (Pearson Correlation, = 0.01). Model development of the bioassay data of the two essential oils allowed estimation of RD_{50} and RD_{75} (Table 2). Previous work at ICIPE, Nairobi, Kenya, tested various DEET doses under the same laboratory conditions as described previously [27, 35, 37] and determined their percent repellence against R. appendiculatus (Table 3), with which we compared the current dose-response repellencies caused by the essential oils of T. minuta and T. diversifolia.

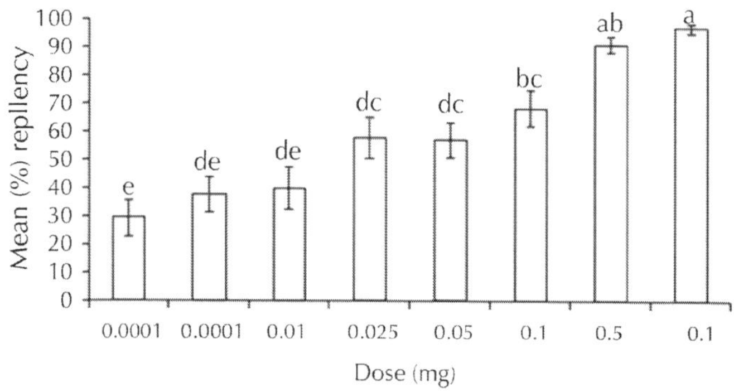

(a)

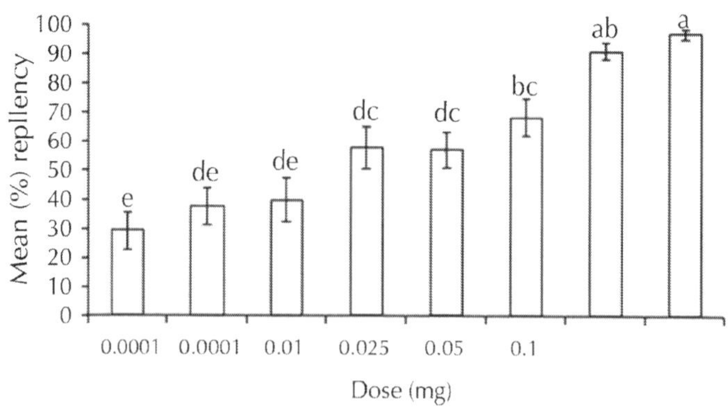

(b)

Figure 2: Mean repellency percentage of different doses of Tagetes minuta (a) andTithonia diversifolia (b) essential oils against newly emerged adults, Rhipicephalus appendiculatus, in a dual-choice assay. Means with the same alphabetical letters are not significantly different at $P<0.0001$ (Student-Newman-Keuls H test).

Major Chemical Components of the Essential Oils of Tagetes Minuta and Tithonia Diversifolia

Gas chromatography (GC) in combination with gas chromatography/ mass spectrometry (GC-MS) separated the chemical components in the mixtures of the essential oils of T. minuta and T. diversifolia plants, and the major representative GC/GC-MS profiles are shown in Tables 1(a) and 1(b), respectively. The major chemical components of T. minuta essential oil were cis-ocimene (43.78%), dihydrotagetone (16.71%), piperitenone (10.15%), trans-tagetone (8.67%), 3.9-epoxy-p-mentha-1.8(10)diene (6.47%), β-ocimene (3.25%), cis-tagetone (1.95%), and β-caryophyllene (0.84%). Those chemical components of the essential oil of T. diversifolia were mainly α-pinene (63.64%), β-pinene (15.00%), isocaryophyllene (7.62%), nerolidol (3.70%), 1-tridecanol (1.75%), limonene (1.52%), and sabinene (1.00%).

Table 1: The major chemical components in the essential oils of Tagetes minuta (a) andTithonia diversifolia (b) identified using capillary column gas chromatography in combination with gas chromatography/mass spectrometry

(a)								
RT (min)	Chemical name	IUPAC name	Molecular formula	M+(g/mol)	Density (g/cm3)	CAS number	Structure	Relative abundance (%)
22.701	cis-Ocimene	3,7-Dimethyl-1,3,7-octatriene	C10H16	136.24	0.800	3338-55-4		43.78
23.126	Dihydrotagetone	2,6-Dimethylocta-7-en-4-one	C10H18O	154.25	0.826	1879-00-1	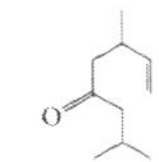	16.71
29.401	Piperitenone	2-Methyl-6-propan-2-ylidenecyclohex-2-en-1-one	C10H14O	150.22	0.977	491-09-8	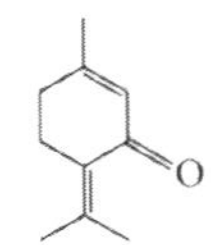	10.15
26.476	trans-Tagetone	(5E)-2,6-dimethylocta-5,7-dien-4-one	C10H16O	152.23	0.847	6752-80-3	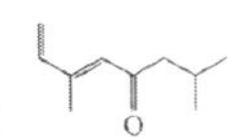	8.67
29.076	3,9 Epoxy-p-mentha-1,8 (10) diene	3,6-Dimethylidene-4,5,7,7a-tetrahydro-3aH-1-benzofuran	C10H14O	150.22	0.967	494-90-6	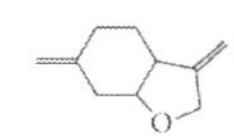	6.47
25.851	β-Ocimene	(3E)-3,7-Dimethylocta-1,3,6-triene	C10H16	136.23	0.776	3779-61-1		3.25
26.126	cis-Tagetone	(Z)-2,6-Dimethylocta-5,7-dien-4-one	C10H16O	152.23	0.847	3588-18-9	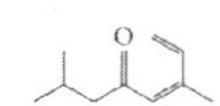	1.95

35.701	β-Caryophyllene (trans)	4,11,11-Trimethyl-8-methylene-bicyclo[7.2.0] undec-4-ene	C15H24	204.36	0.905	87-44-5	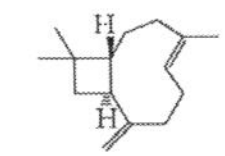	0.84
37.726	Bicyclogermacrene	(1R, 4E,8E, 10S)-4,8,11,11-Tetramethylbicyclo[8.1.0] undeca-4,8-diene	C15H24	204.35	0.861	24703-35-3	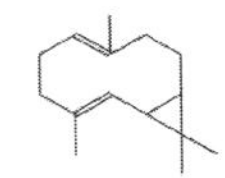	0.62
50.626	AR-turmerone	(6S)-2-Methyl-6-(4-methylphenyl)hept-2-en-4-one	C15H20O	216.32	0.945	532-65-0	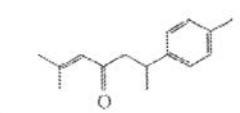	0.50

Key:

M⁺: molecular weight.

RT: retention time in minutes (min.).

IUPAC: The IUPAC nomenclature system of organic chemistry is a systematic method of naming organic chemical compounds as recommended by the International Union of Pure and Applied Chemistry (IUPAC).

CAS registry numbers: Are unique numerical identifier (values) numbers assigned by the chemical abstracts service (CAS) to every chemical described in the open scientific literature (currently including those chemicals described from at least 1957 through the present) and including elements, isotopes, organic and inorganic compounds, ions, organometallics, metals, nonstructurable materials, and so forth.

Table 2: Probit analysis of dose-response relationship of Tagetes minuta and Tithonia diversifolia essential oils at RD_{50} and RD_{75} generated by a regression model: Probit [Π(dose1)] = 1.1036 + 0.4132 dose1 for the essential oil of T. minuta and the regression model: Probit [Π(dose1)] = 0.6401 + 0.4962 dose1 for the essential oil of T. diversifolia

Plant species	Repellence probability	Repellent dose (mg)	Upper confidence limit at 95%	Lower confidence limit at 95%
Tagetes minuta	0.50	0.0021	0.0024	0.0019
	0.75	0.0915	0.1012	0.0830
Tithonia diversifolia	0.50	0.2629	0.2712	0.2548
	0.75	0.5972	0.6116	0.5835

Table 3: Mean percent repellence (±SE) of N,N-diethyl-3-methylbenzamide (DEET) evaluated in a dual-choice assay against newly emerged adult ticks, Rhipicephalus appendiculatus, at the International Centre of Insect Physiology and Ecology, Nairobi, Kenya, under the same laboratory conditions as the current studies

Repellent dose (mg)	Repellency (%)
0.0998	84.0± 3.9
0.00998	82.8± 3.6
0.000998	75.6± 4.5
0.0000998	70.5± 3.6

Sources: Ndung'u et al. [35], 1995; Lwande et al. [37], 1998; Ndung'u et al. [27], 1999.

DISCUSSION

In a previous study, we undertook a survey of indigenous knowledge of the Bukusu community of western Kenya on livestock ticks, the risks they pose and ethnopractices associated with their management [41]. The study showed that the Bukusu community has accumulated rich ethnoveterinary knowledge and practices and that on-host use of ethnobotanical suspensions and fumigation of host animals with volatiles from burning plant products (prepared from one or more of ~157 plants) constitute important methods of controlling the ticks. In the present study, repellence of essential oils associated with some of the plants was used to assess one possible mode of action of the plant products on ticks. Eight of these plants were selected for screening against the adults of the brown ear tick following a four-level assessment protocol (based on additional ethnobotanical information on similar use of the plant elsewhere, reported phytochemical profile of the plant or related species, and any documented bioactivity data of the plant extracts or their phytochemicals) proposed by Heinrich and coworkers [42]. The essential oils of these plants showed some variation in repellence against newly emerged R. appendiculatus adult ticks. This was particularly apparent at the lower dose (0.1 mg), with the essential oil of T. minuta showing the highest repellence and that of S. longepedunculata showing the least repellent effect.

Comprehensive repellence studies with T. minuta and T. diversifolia oils at eight doses confirmed the higher repellence of the former against R. appendiculatus. Interestingly, its repellent effect is comparable to that of commonly used repellent DEET at 0.1 mg dose (with essential oil of T. minuta producing a repellent effect of 80.1 ± 4.9% compared with that of DEED, 84.0 ± 3.9%). Additionally, the results obtained by this study are compared favourably with the results reported by Nchu and coworkers [52] for the essential oil of T. minutaplants growing in Pretoria, South Africa, against Hyalomma rufipes Koch. H. rufipes is also widely distributed in Africa and can transmit disease-causing viral and protozoan (e.g., Crimean-Congo haemorrhagic fever and Babesia, resp.) pathogens to livestock and humans alike [53, 54]. Although the T. minuta oil obtained in the present study shares a series of major constituents (e.g., cis-ocimene, dihydrotagetone, piperitenone, tagetone, and -ocimene) with that isolated in Pretoria, South Africa [52], there are also some chemotypic differences between the two, reflected in different proportions of these compounds and the presence of some constituents in one chemotype (e.g., 3-methyl-2-(2-methyl-2-butenyl)-furan in T. minuta oil collected in Pretoria, South Africa) that were not detected in the other. Which constituents contribute to the repellence of the essential oil of T. minuta against the two tick species, respectively, must await detailed subtractive assays [55] with synthetic blends of the major constituents of the two essential oils with each component missing at a time.

Nchu and coworkers [52] also found that T. minuta oil significantly delayed moulting of H. rufipes engorged nymphs. In another study reported by Krishna and coworkers [56], most eggs of the coleopteran beetle,Tribolium castaneum, exposed to the vapours of essential oil of a specific genotype of T. minuta failed to hatch. Both these effects have been attributed to tagetone, one of the major constituents of T. minutapreviously shown to have growth disrupting bioactivities on arthropods [57]. Since trans-tagetone is a prominent constituent of T. minuta essential oil of Bungoma chemotype (8.7%, compared with 1.6% in the Pretoria chemotype), it will be interesting to see if similar repellent effects are also observed with nymphal R. appendiculatus.

In this study, essential oil of T. diversifolia growing in Bumgoma was found to be less repellent than that of T. minuta. This is reflected in its very different terpenoid profile. However, the plant is highly valued for its tick control property by the Bukusu community in Bungoma [41]. Interestingly, it is also used by the Kikuyu community in central Kenya to control livestock ticks [58] and by the Luyha and Luo communities in western Kenya to control insect pests in arable farming systems [59, 60]. This suggests other possible modes of action of T. diversifolia phytochemical profile. Follow-up studies on other potential modes of action of the essential oil and nonvolatile constituents of the plant may help to shed light on this question.

CONCLUSIONS

In conclusion, the present study provides some scientific rationale for the incorporation of some botanicals in Bukusu ethnopractices in western Kenya to control tick infestations on cattle. The study also lays down some groundwork for follow-up studies on other possible effects of the phytochemicals of the plants studied and for exploiting partially refined products such as essential oils in protecting cattle against infestations by R. appendiculatus and other tick species.

ACKNOWLEDGMENTS

The authors wish to thank Mr. Simon M. Mathenge from the University of Nairobi Herbarium for identification of plant species; ICIPE's staff for offering their technical and administrative support, and Bukusu community ethnopractitioners and livestock farmers for their cooperation, participation and support throughout the study period. They also wish to thank the local administration of the Government of Kenya for supporting, participating, and giving permission to sample plants for isolation of essential oil.

REFERENCES

1. R. A. I. Norval, B. D. Perry, and A. S. Young, The Epidemiology of Theileriosis in Africa, Academic Press, London, UK, 1992.
2. J. M. Olwoch, B. Reyers, F. A. Engelbrecht, and B. F. N. Erasmus, "Climate change and the tick-borne disease, Theileriosis (East Coast fever) in sub-Saharan Africa," Journal of Arid Environments, vol. 72, no. 2, pp. 108–120, 2008.
3. A. J. Lawrence, J. A. de Vos, and A. D. Irvin, "East coast fever," in Infectious Diseases of LivesTock With Special Reference to Southern Africa, G. R. Thomson and R. C. Tustin, Eds., Cape Town Oxford University Press, Cape Town, South Africa, 1988.
4. K. S. Mbogo, P. D. Kariuki, N. McHardy, and R. Payne, Training Manual For Veterinary Staff on Immunization Against East Coast Fever Using the ECFiM System, Kenya Agricultural Research Institute and Overseas Development Administration of UK, Nairobi, Kenya, 1995.
5. "Future of livestock industries in East and southern Africa," in Proceedings of the a Workshop Held at Kadoma Ranch Hotel, Zimbabwe International Livestock Centre for Africa (ILCA '92), J. A. Kategile and S. Mubi, Eds., pp. 20–23, Addis Ababa, Ethiopia, July 1992.
6. A. Hassanali, H. Herren, Z. R. Khan, J. A. Pickett, and C. M. Woodcock, "Integrated pest management: the push-pull approach for controlling insect pests and weeds of cereals, and its potential for other agricultural systems including animal

husbandry," Philosophical Transactions of the Royal Society B: Biological Sciences, vol. 363, no. 1491, pp. 611–621, 2008.

7. S. M. Cook, Z. R. Khan, and J. A. Pickett, "The use of push-pull strategies in integrated pest management," Annual Review of Entomology, vol. 52, pp. 375–400, 2007.
8. B. Fei and J. H. Xin, "N, N-diethyl-m-toluamide-containing microcapsules for bio-cloth finishing,"American Journal of Tropical Medicine and Hygiene, vol. 77, no. 1, pp. 52–57, 2007.
9. A. T. Mathai, K. S. Pillai, and P. B. Deshmukh, "Acute toxicity of deet to a freshwater fish, Tilapia mossambica: effect on tissue glutathione levels," Journal of Environmental Biology, vol. 10, no. 2, pp. 87–91, 1989.
10. J. Seo, Y.-G. Lee, S.-D. Kim, C.-J. Cha, J.-H. Ahn, and H.-G. Hur, "Biodegradation of the insecticideN,N-diethyl-m-toluamide by fungi: identification and toxicity of metabolites," Archives of Environmental Contamination and Toxicology, vol. 48, no. 3, pp. 323–328, 2005.
11. P. J. Robbins and M. G. Cherniack, "Review of the biodistribution and toxicity of the insect repellentN,N-diethyl-m-toluamide (DEET)," Journal of Toxicology and Environmental Health, vol. 18, no. 4, pp. 503–525, 1986.
12. H. Qiu, H. W. Jun, and J. W. Mccall, "Pharmacokinetics, formulation, and safety of insect repellent N,N-Diethyl-3-methylbenzamide (deet): a review," Journal of the American Mosquito Control Association, vol. 14, no. 1, pp. 12–27, 1998.
13. M. C. Lewis, M. Silva, J. Sanborn, K. Pfeifer, P. J. Schreider, and E. P. Helliker, N, N-Diethyl-Meta-Toluamide (DEET): Risk Characterization Document, Medical Toxicology and Worker Health and Safety Branches: California Environmental Protection Agency, Department of Pesticide Regulation. Pesticide Registration Branch, Sacramento, Calif, USA, 2000,http://www.cdpr.ca.gov/docs/risk/rcd/deet.pdf.
14. R. Menendenz, "El Melinitus minutifloraay la garrapata," Agricultural Review, vol. 4, pp. 219–223, 1924.
15. K. C. Thompson, J. Roa, and T. Romero, "Anti-tick grasses as the basis for developing practical tropical tick control packages,"

Tropical Animal Health and Production, vol. 10, no. 3, pp. 179–182, 1978

16. J. F. Carroll, A. Maradufu, and J. D. Warthen Jr., "An extract of Commiphora erythraea: a repellent and toxicant against ticks," Entomologia Experimentalis et Applicata, vol. 53, no. 2, pp. 111–116, 1989.
17. L. J. Wilson and R. W. Sutherst, "Oviposition sites of Boophilus microplus Canestrini (Acarina: Ixodidae) in Stylosanthes and grass pastures," vol. 29, no. 2, pp. 101–105, 1990.
18. E. C. Webb and M. David, "The efficacy of neem seed extract (Azadirachta indica) to control tick infestation in Tswana, Simmentaler and Brahman cattle," South African Journal of Animal Sciences, vol. 32, no. 1, pp. 1–6, 2002.
19. N. R. Lersten, "Crystals of calcium compounds in Gramineae," New Phytologist, vol. 93, no. 4, pp. 633–638, 1983.
20. F. I. J. Lagunes, D. G. Fox, R. W. Blake, and A. N. Pell, "Evaluation of tropical grasses for milk production by dual-purpose cows in Tropical Mexico," Journal of Dairy Science, vol. 82, no. 10, pp. 2136–2145, 1999.
21. J. A. Duke, "Duke's handbook of energy crops (unpublished)," 1983,http://www.hort.purdue.edu/newcrop/duke_energy/Melinis_minutiflora.html.
22. J. W. Chiera, R. M. Newson, and A. Maradufu, "Anti-tick properties of molasses grass, Melinis minutiflora," in Proceedings of the 5th annual report ICIPE, ICIPE Press, Nairobi, Kenya, 1977.
23. L. Hernandez, D. Parra, and A. Ahumada, "Repellent and acaricidal activity of the oil and some chromatographic fractions of the grass Melinis minutiflora against Boophilus microplus," Revista Colombiana de Ciencias Químico Farmacéuticas, vol. 17, pp. 45–50, 1990.
24. E. N. Mwangi, S. Essuman, G. P. Kaaya, E. Nyandat, D. Munyinyi, and M. G. Kimondo, "Repellence of the tick Rhipicephalus appendiculatus by the grass Melinis minutiflora," Tropical Animal Health and Production, vol. 27, no. 4, pp. 211–216, 1995.
25. O. O. Dipeolu, O. A. Mongi, K. D. Punya, A. A. Latif, A. O. Amoo, and T. R. Odhiambo, "Current concepts and approach to control

of livestock ticks in Africa," Discovery and Innovation, vol. 4, no. 2, pp. 35–44, 1992.

26. M. M. Malonza, O. O. Dipeolu, A. O. Amoo, and S. M. Hassan, "Laboratory and field observations on anti-tick properties of the plant Gynandropsis gynandra (L.) Brig," Veterinary Parasitology, vol. 42, no. 1-2, pp. 123–136, 1992.
27. M. W. Ndung'u, S. C. Chhabra, and W. Lwande, "Cleome hirta essential oil as livestock tick (Rhipicephalus appendiculatus) and maize weevil (Sitophilus zeamais) repellent," Fitoterapia, vol. 70, no. 5, pp. 514–516, 1999.
28. M. Jacobson, Plants: the Potentials For Extracting Protein, Medicines and Other Useful Chemicals: Workshop Proceedings, Office of Technology Assessment, Washington, DC, USA, 1983.
29. H. Panda, Aromatic Plants: Cultivation, Processing and Uses, Asia Pacific Business Press, New Delhi, India, 2004.
30. Y. Trongtokit, Y. Rongsriyam, N. Komalamisra, P. Krisadaphong, and C. Apiwathnasorn, "Laboratory and field trial of developing medicinal local thai plant products against four species of mosquito vectors," Southeast Asian Journal of Tropical Medicine and Public Health, vol. 35, no. 2, pp. 325–333, 2004.
31. K. R. Chauhan, J. A. Klun, M. Debboun, and M. Kramer, "Feeding deterrent effects of catnip oil components compared with two synthetic amides against Aedes aegypti," Journal of Medical Entomology, vol. 42, no. 4, pp. 643–646, 2005.
32. B. A. Bond, Herbal Tick Repellent for Dogs, Care 2 Green Living Newsletters,2007,http://www.care2.com/greenliving/tick-repellant.html.
33. Nixalite of America, Flea and Tick Granular Repellent: Safe, All Natural, Biodegradable, ABC/Nixalite, Nixalite of America, East Moline, Ill, USA, 2005.
34. A. Maradufu, "Furanosesquiterpenoids of Commiphora erythraea and C. myrrh," Phytochemistry, vol. 21, no. 3, pp. 677–680, 1982.
35. M. W. Ndung'u, W. Lwande, A. Hassanali, L. Moreka, and S. C. Chhabra, "Cleome monophylla essential oil and its constituents as tick (Rhipicephalus appendiculatus) and maize weevil (Sitophilus zeamais) repellents," Entomologia Experimentalis et Applicata, vol. 76, no. 3, pp. 217–222, 1995.

36. E. N. Mwangi, A. Hassanali, S. Essuman, E. Myandat, L. Moreka, and M. Kimondo, "Repellent and acaricidal properties of Ocimum suave against Rhipicephalus appendiculatus ticks," Experimental and Applied Acarology, vol. 19, no. 1, pp. 11–18, 1995.
37. W. Lwande, A. J. Ndakala, A. Hassanali et al., "Gynandropsis gynandra essential oil and its constituents as tick (Rhipicephalus appendiculatus) repellents," Phytochemistry, vol. 50, no. 3, pp. 401–405, 1998
38. R. K. Gupta and L. C. Rutledge, "Role of repellents in vector control and disease prevention," American Journal of Tropical Medicine and Hygiene, vol. 50, no. 6, pp. 82–86, 1994.
39. R. A. Copeland, T. W. Walker, L. L. Robert, J. I. Githure, R. A. Wirtz, and T. A. Klein, "Response of wildAnopheles funestus to repellent-protected volunteers is unaffected by malaria infection of the vector,"Journal of the American Mosquito Control Association, vol. 11, no. 4, pp. 438–440, 1995.
40. WHO, International Travel and Health 1995: Vaccination Requirements and Health Advice (Vaccination Certificate Requirements For International Travel and Health Advice To Travellers), World Health Organization, Geneva, Switzerland, 1995.
41. W. Wanzala, W. Takken, W. R. Mukabana, A. O. Pala, and A. Hassanali, "Ethnoknowledge of Bukusu community on livestock tick prevention and control in Bungoma district, western Kenya," Journal of Ethnopharmacology, vol. 140, no. 2, pp. 298–324, 2012.
42. M. Heinrich, H. Rimpler, and N. A. Barrera, "Indigenous phytotherapy of gastrointestinal disorders in a lowland Mixe community (Oaxaca, Mexico): ethnopharmacologic evaluation," Journal of Ethnopharmacology, vol. 36, no. 1, pp. 63–80, 1992.
43. W. Wanzala, Ethnobotanicals For Management of the Brown Ear Tick, Rhipicephalus Appendiculatus in Western Kenya, Ponsen and Looijen, Wageningen, The Netherlands, 2009.
44. K. P. Bailey, "Notes on the rearing of Rhipicephalus appendiculatus and their infection with Theileria parva for experimental transmission," Bulletin of epizootic diseases of Africa, vol. 8, pp. 33–43, 1960.

45. A. D. Irvin and D. W. Brocklesby, "Rearing and maintaining Rhipicephalus appendiculatus in the laboratory," Ji Nan Idoon Animation Technology, vol. 21, pp. 106–112, 1970.
46. J. F. Clevenger, "Apparatus for the determination of volatile oil," Journal of the American Pharmaceutical Association, vol. 17, no. 4, pp. 345–349, 1928.
47. W. Wanzala, N. F. K. Sika, S. Gule, and A. Hassanali, "Attractive and repellent host odours guide ticks to their respective feeding sites," Chemoecology, vol. 14, no. 3-4, pp. 229–232, 2004.
48. T. O. Browning, "The aggregation of questing ticks, Rhipicephalus pulchellus, on grass stems, with observations on Rhipicephalus appendicuiatus," Physiological Entomology, vol. 1, no. 2, pp. 107–114, 1976.
49. J. W. Chiera, "The distribution and climbing behaviour of Rhipicephalus appendiculatus Neumann on grass stems," Insect Science, vol. 6, no. 2, pp. 213–215, 1985.
50. SAS Institute Inc, Proprietary Software Release 8. 1 (TS1M0), SAS Institute, SAS Campus Drive, Cary, NC, USA, 2003.
51. C. I. Bliss, "The method of probits," Science, vol. 79, no. 2037, pp. 38–39, 1934.
52. F. Nchu, S. R. Magano, and J. N. Eloff, "In vitro anti-tick properties of the essential oil of Tagetes minutaL. (Asteraceae) on Hyalomma rufipes (Acari: Ixodidae)," Onderstepoort Journal of Veterinary Research, vol. 79, no. 1, pp. 1–5, 2012.
53. E. H. Graven, L. Webber, G. Benians, M. Venter, and J. B. Gardner, "Effect of soil type and nutrient status on the yield and composition of Tagetes oil (Tagetes minuta L.)," Journal of Essential Oil Research, vol. 3, no. 5, pp. 303–307, 1991.
54. R. A. Walker, A. Bouattour, J. L. Camicas et al., Ticks of Domestic Animals in Africa: A Guide To Identification of Species. Bioscience Reports, University of Edinburgh, Edinburgh, UK, 2003.
55. J. Bekele and A. Hassanali, "Blend effects in the toxicity of the essential oil constituents of Ocimum kilimandscharicum and Ocimum kenyense (Labiateae) on two post-harvest insect pests,"Phytochemistry, vol. 57, no.a 3, pp. 385–391, 2001.
56. A. Krishna, V. Prajapati, S. Bhasney, A. K. Tripathi, and S. Kumar, "Potential toxicity of new genotypes of Tagetes (Asteraceae)

species against stored grain insect pests," International Journal of Tropical Insect Science, vol. 25, no. 2, pp. 122–128, 2005.

57. C. O. Adedire and J. O. Akinneye, "Biological activity of tree marigold, Tithonia diversifolia, on cowpea seed bruchid, Callosobruchus maculates (Coleoptera: Bruchidae)," Annals of Applied Biology, vol. 144, no. 2, pp. 185–189, 2004.

58. G. N. Njoroge and R. W. Bussmann, "Herbal usage and informant consensus in ethnoveterinary management of cattle diseases among the Kikuyus (Central Kenya)," Journal of Ethnopharmacology, vol. 108, no. 3, pp. 332–339, 2006.

59. F. Adoyo, B. J. Mukalama, and M. Enyola, "Using Tithonia concoctions for termite control in Busia District, Kenya: As part of the Kenya Woodfuel and Agroforestry Programme (KWAP) farmers in the Busia District of western Kenya taking part in an on-farm experiment in pest control. Research and Information Exchange in EcologicallySound Agriculture (ILEIA)," News Letter, vol. 13, no. 4, pp. 24–25, 1997.

60. Anonymous, Termite Control. HDRA-the Organic Organization. Ryton Organic Gardens, COVENTRY CV8 3LG, 2001, www.hdra.org.uk.

Chapter 10

Efficacy of Oryza Sativa Husk and Quercus Phillyraeoidesextracts for the in vitro and in vivo Control of Fungal Rot Disease of White Yam (Dioscorea Rotundata Poir)

Victor Ohileobo Dania[1,2], Olubunmi Omowunmi Fadina[2], Maria Ayodele[1], and P Lava Kumar[1]

[1]International Institute of Tropical Agriculture (IITA), Ibadan, Nigeria

[2]Department of Crop Protection and Environmental Biology, University of Ibadan, Ibadan, Nigeria

ABSTRACT

Tuber rot disease is a major constraint to white yam (*Dioscorea rotundata*) production, accounting for 50-60% of annual yield losses in Nigeria. The main method of control using synthetic fungicides is being discouraged due to human and environmental health hazards. The potential of *Oryza sativa* husk (OSH) and *Quercus phillyraeoides* (QP) extracts for the *in vitro* and *in vivo* control of six virulent rot-causing fungal pathogens, *Lasiodiplodia theobromae, Aspergillus niger, Rhizoctonia solani, Penicillium oxalicum, Sclerotium rolfsii,* and *Fusarium oxysporum* was evaluated, using five different extract concentrations of 0.5%, 1.0%, 1.5%, 2.5%, and 3.5% w/v. These fungi were isolated from rotted tubers of *D. rotundata,* across three agroecological zones in Nigeria—the Humid rainforest, Derived savanna, and southern Guinea savanna. All treatments were subjected to three methods of inoculation 48 hours before the application of both extracts and stored at 28±2°C for 6 months. Radial mycelial growth of the test pathogens was effectively inhibited at concentrations≤3.5% w/v *in vitro* for both OSH and QP extracts. Rotting was significantly reduced (P≤0.05) to between 0 to 18.8% and 0% to 20.9% for OSH and QP extracts respectively. The extracts significantly (P≤0.05) inhibited percent rot of the test pathogens at 3.5% concentration w/v *in vivo*. Rot incidence was, however, lower in replicate tubers that were inoculated, treated with extracts and exposed than treatments that were covered. Phytochemical analysis of OSH and QP extracts revealed the presence of secondary metabolites such as alkaloids, flavonoids, saponins, tannins, ferulic acid, phlobatanins, Terpenoids, phenols, anthraquinone and pyroligneous acid. The efficacy of both extracts in reducing rot in this study recommends their development as prospective biopesticide formulation and use in the management of post-harvest rot of yam tubers.

INTRODUCTION

Yam is mainly cultivated as a staple crop across the humid and semi-humid tropics, especially in West Africa which accounts for about 90% of world yam production (Aboagyne-Nuamah et al.2005). There are over 150 species of yam grown throughout the world (Purseglove 1985). White yam (*D. rotundata*) is the most predominant yam species and it constitutes about 80% of total yam produced worldwide [(Titteh and Saakwa 1994; Ejechi and Souzey 1999). Although the main constituent of the yam tuber is carbohydrate, it also contains an appreciable amount of proteins, vitamins and minerals (Babaleye 2003). It serves as an important source of income to many people. The tuber is of economic value and is the most important part of the crop. It can be eaten in different forms such as roasted, boiled, fried or better still made as pounded yam or as flour which is a delicacy in Nigeria. Yam is a very important and revered crop, especially in southern Nigeria where it has cultural importance exemplified by the annual new yam festival (Arua 1981; Okorji 1992). Water yam (*D. alata*) becomes an alternative only when the white yam is out of season or very costly.

Pests and diseases are the major limitation to cultivation; hence demand has always exceeded its supply in yam producting countries like Nigeria (Coursey 1983). Yam tubers are liable to infection by pathogens such as fungi, bacteria, viruses, and nematodes at various stages of its growth in the field. In most cases, infection starts in the field through injuries to the yam tubers and in transit to the market or barn for storage (Yoshida et al. 2000). Rot is an important factor affecting post-harvest shelf life of yams and losses can be very high. Losses of over 60% of *D. rotundata* tubers rot stored for six months have been reported (Ikotun 1983a, 1989). The yield loss ultimately reduces farmers' income, available food supply and yam setts for the next growing season. (Amusa et al. 2003; Aidoo 2007). Besides the decrease in quantity, rot also reduces the quality of yam tubers thereby making them unattractive to prospective buyers. Yams are attacked by several diseases; fungi are, however, the major cause of post-harvest rot of yam tubers (Noon1978; Dania 2012). Fungal pathogens such as *Lasiodiplodia theobromae, Aspergillus niger,Fusarium oxysporum, Rhizoctonia solani, Penicillium oxalicum,* and *Sclerotium rolfsii* are commonly associated with rotting of yam tubers (Ikotun 1989; Aidoo 2007).

Peasant farmers often quickly sell off their harvested yam tubers in order to avoid losses due to microbial decay. This action has adversely affected yam production, availability and utilization. Yam producers are always concerned with tuber quality from cultivation on the farm to storage in order to guarantee high price for their produce. The quality of yam setts for the next growing season is also very important to producers in order to ensure perpetuation of the crop and guarantee continuous supply. This, therefore, emphasizes the need for disease control.

Synthetic fungicides had been very popular in the control of agricultural pests and pathogens because of their ease of application, effectiveness and storage (Prakash and Rao 1994). However, in recent times, their use in crop protection programmes has come under increased criticisms worldwide for some obvious reasons. Their use results in adverse effects on the environment, persistence, pathogen resurgence due to resistance development, mammalian toxicity, elimination of beneficial soil micro and macro-organisms (GOAN 1999). There is, therefore, a need to search for cheap and eco-friendly control measures in the management of post-harvest rot of yam. Several authors have suggested the use of botanicals as an alternative to the hazardous effects of synthetic pesticides (Enikuomehin et al. 1998; Amadioha 2000; Hycenth 2008). However, most of the efforts with botanicals to control yam pathogens do not go beyond the *in vitro* trials. Hence this study investigated the potential of *Oryza sativa* husk and *Quercus phillyraeoides* extracts for the control of post-harvest fungal rot and extension of the shelf life of white yam.

MATERIALS AND METHODS

Collection and Isolation of Associated Fungi from Rotted Yam Tuber Samples

Rotted tuber samples of *D. rotundata* were collected from different locations across three agroecological zones (AEZ): Humid rainforest, Derived savanna and southern Guinea savanna. Fifteen tubers showing rot symptoms were collected from fourteen locations across the three AEZ. A total of 210 rotten tubers were collected and evaluated for fungal rot pathogens. Each rotten sample was washed in running tap water and cut to expose the fresh necrotic tissues. From the areas in advance of the necrosis, tissue pieces were surface-sterilized with 10% sodium hypochlorite for 2 minutes, and rinsed in five changes of sterile distilled water (SDW) before plating on potato dextrose agar (PDA). The inoculated plates were incubated in the dark in a Gallenkamp incubator operating at a temperature of 28°C for 3 to 4 days and examined for fungal growth from the tissue pieces. Identification of the isolates was carried out using standard procedure (Barnett and Hunter 1998).

Preparation of Inoculum

Pure cultures of the pathogens were subcultured from different stock cultures. These were placed on PDA for 7–10 days depending on the pathogen, to allow full sporulation. This was then scooped out into a Warring blender containing one litre of SDW; a drop of Tween 20 (Polyoxyethylene sorbitan mono-oleat) detergent was added to aid the dislodgment of the spores from the medium. The spores were then strained off the mycelia fragments using a double layer of cheese cloth placed inside a sterile funnel. An aliquot of 0.1 mL spore concentration was then placed on an hemacytometer slide and the spores were quantified under a compound microscope. The final spore count and concentration was determined by using this method:

$$C_1V_1=C_2V_2$$

Where: C_1=Initial inoculum concentration, V_1=Initial volume of water used in streaking the culture plate

C_2=Final inoculum concentration desired, V_2=Final volume of water added to obtain desired concentration.

Preparation of Cold Water Extracts

Fresh *Oryza sativa* husks were washed thoroughly under running tap water and air-dried at a temperature of 28°C for 7–10 days until they became crisp. The dried plant husks were pulverized by blending using a high speed blender (Waring Commercial, Springfield, MO, USA) to form a fine powder while the bark of *Quercus phillyraeodes* (Oak plant) was dried in a Gallenkamp oven at 80°C for two days and crushed to a coarse powder in a mortar. Ultraviolet light (UV) was used to surface-sterilize the botanicals to prevent fungal and bacterial contaminants. Cold water extract was prepared by adding the powder to 100 mL of sterile distilled water in a 250- mL beaker. This was stirred vigorously and allowed to stand for 24 hours before filtering the extract through folds of sterile cheese cloth. Five different extract concentrations were prepared by blending 0.5 g, 1.0 g, 1.5 g, 2.5 g, and 3.5 g of each plant part in 100 mL of sterile distilled water to produce 0.5%, 1.0%, 1.5%, 2.5%, and 3.5% extract concentrations, respectively.

Measurement of Mycelial Growth Inhibition

This involved creating a four equidistant section on each Petri dish by drawing two perpendicular lines at the reverse bottom of the plate, the point of intersection indicating the center of the plate. This was done before dispensing PDA into each of the plates. An aliquot of 1 mL each of the extracts was separately introduced into the Petri dish containing 15mL PDA. A 5-mm cork borer was used to inoculate a disc of the pathogen culture and placed on the medium containing extract just at the point of intersection of the two previously drawn lines at the bottom of the Petri dish in three replicates. Control experiments were set up without the addition of any plant material. Inhibitory effect of the extract was expressed as percentage inhibition and calculated using the formula:

$$\%\text{Mycelialinhibition} = [(XC-YT)/XC] \times 100$$

Where: XC=Average diameter of control

YT=Average diameter of fungal colony with treatment

Determination of the Efficacy of Extracts for Control of Tuber Rot

The experiment was a 2×3×6 factorial in a randomized complete block design with three replicates. Both extracts that had earlier proved effective in the *in vitro* inhibition of the pathogens were used for the trials against the virulent fungal pathogens. Clean, healthy tubers of yam genotype TDr 95–18544 were washed under running tap water to remove adhering soil and surface-borne microflora. Eighteen tubers were inoculated per pathogen with three replicates in two batches. The first batch of nine tubers consisting of three sets of tubers in each treatment, were separately scratched, wounded, and bored, respectively. The tubers were inoculated with the test pathogens, the surface area treated with 1 ml of the extract and then exposed. Similarly, the second batch of nine tubers was scratched, wounded, bored, prior to inoculation and treated with the extract and covered with black polyethylene sheets. For scratched treatments, the tuber periderm was removed with a scapel up to 0.1 cm; wounded tubers were cut to a depth of 0.5 cm. Bored treatments were inoculated to a depth of 1 cm, using a 5-mm cork borer. The tubers were inoculated with the test pathogens 48 h before the application of the extracts and stored at room temperature (28±2°C) in the yam barn for 6 months. Tubers were inoculated with a spore concentration of 10^6 conidia/mL of the test pathogen. Control tubers were scratched, bored, wounded, and dipped in SDW and inoculated with test pathogen only. Percent rot inhibition was determined using the method:

$$\text{Rot inhibition} = \frac{\%\,\text{decay in contraol} - \%\,\text{decay in treated tuber}}{\%\,\text{decay in control}}$$

Phytochemical Screening of Plant Extracts

Fresh samples of *O. sativa* husk and *Q. phillyraeoides* bark extracts were prepared as previously described and were concentrated using a rotary evaporator. Chemical tests were carried out on the extracts for the qualitative determination of phytochemical constituents using standard procedures (Sofowora 1993). Mayer's and Draggendoff's reagents were used for test and confirmation of the presence of alkaloids in the extracts. The formation of a cream colour and reddish brown precipitate with Mayer's and Draggendoff's reagent respectively was regarded as the confirmatory test for the presence of alkaloids. The amount of total phenolics in extracts was determined with the Folin-Ciocalteu reagent. Gallic acid was used as a standard and the total phenolics were expressed as mg/g gallic acid equivalents (GAE). The formation of blue colour upon reaction with Folin-Ciocalteu reagent was the confirmatory test for the presence of phenols in the plant extract samples. Similarly, the formation of a red precipitate when the extract was boiled with 1% aqueous hydrochloric acid indicated the presence of phlobatannins. Other bioactive constituents such as saponins, tanins, flavonoids, steroids, terpenoides, anthraquinones etc. were also determined using standard procedures.

Data Analysis

All numerical data were statistically analysed using generalized linear model (GLM) of SAS. Means were separated using Least Significant Difference Test (LSD) and standard error was determined where applicable at 5% level of significance.

RESULTS

Mycelial Growth Inhibition of Test Pathogens *in Vitro*

Mycelial growth inhibition of the test pathogens ranged between 0% to 39.2% for OSH and 0% to 46.6% for QP extracts at 0.5% w/v concentration (Table 1). However, both extracts effectively inhibited growth of the pathogens at 3.5% w/v concentration (>70% inhibition). The result also shows that OSH could not completely inhibit *A.niger in vitro* at concentrations≤2.5% w/v. There was no significant difference (P>0.05) in the control *of L. theobromae and A. niger* at 0.5% w/v concentration for both OSH and QP extracts. However, the efficacy of both extracts differed significantly (P≤0.05) at concentrations ≥1.0%≤2.5% w/v. There was no significant difference in the efficacy of OSH and QP against *S. rolfsii* and *F. oxysporum* at 3.5% w/v concentration. The minimum inhibitory concentration for both extracts varied between 0.5% to 1.5% w/v depending on the test pathogen and extract.

Table 1: Effect of *Oryza sativa* husk and *Quercus phillyraeoides* extracts on the mycelial growth of six virulent fungi after 7 days

Mycelial growth inhibition						
	Extract	0.5	1.0	1.5	2.5	3.5
L. theobromae	OSH	0.0e	100a	100a	100a	100a
	QP	0.0e	0.0e	11.1e	62.2 cd	86.1bc

A. niger	OSH	5.6 cd	11.1cde	70.0bc	77.8c	100a
	QP	6.7fcd	22.3c	33.3d	61.7 cd	77.4c
S. rolfsii	OSH	12.2c	22.2c	55.9 cd	89.2b	100a
	QP	0.0e	17.3 cd	65.4c	100a	100a
P. oxalicum	OSH	23.1abc	24.4c	67.2c	83.1bc	100a
	QP	30.0bc	52.3abc	100a	100a	100a
R. solani	OSH	4.4 cd	100a	100a	100a	100a
	QP	0.0e	0.0e	16.4de	88.0b	93.8b
F. oxysporum	OSH	39.2b	63.2bc	100a	100a	100a
	QP	46.6a	70.6b	90.6b	100a	100a
Std deviation		16.2	36.0	32.8	14.6	7.3

Means with the same letter along a column are not significantly different (P=0.05).

Legend: OSH=Oryza sativa husk, QP=Quercus phillyraeoides.

Dania et al.

Dania et al. SpringerPlus 2014 3:711, doi: 10.1186/2193-1801-3-711

Efficacy of Extracts for Control of Tuber Rot

Rot development was significantly reduced (P≤0.05) in tubers of *D. rotundata* inoculated with six virulent fungal pathogens 48 h before the application of *O. sativa* husk extract (Table 2). Rot initiation in replicate tubers that were scratched, inoculated with *R. solani, P. oxalicum* and *F. oxysporum* and treated with *O.sativa* extract was completely inhibited. Rot inhibition was more effective in inoculated tubers that were either scratched or wounded and treated with the extracts than the bored treatments. However, the extract was more effective in tubers that were bored than those that were wounded, inoculated with *L. theobromae* and treated with the extract where rot was reduced to 17.4% and 18.8% respectively. Comparatively, rot was very pronounced in control treatments that were either wounded or bored and inoculated with the same pathogen but without extract application (52.2% to 78.2% rot). There was significant difference (P≤0.05) in rot incidence between all the extract applied treatments and the control tubers that were inoculated with the test pathogens only.

Table 2: Rot inhibition in tubers inoculated with six virulent fungi 48 h before the application of *O. sativa* extract stored for 6 months

Tuber rot (%)						
Fungus	Inoculated and exposed			Inoculated and covered		
	Scratch	Wound	Bore	Scratch	Wound	Bore
L. theobromae	5.5bc	10.4 cd	12.0d	18.8de	18.8d	17.4d
Control	12.8a	15.8bc	33.7bc	17.6bc	52.2a	78.2b
R. solani	0.0 cd	5.6de	11.8d	7.9e	11.2de	15.9de
Control	10.2b	9.5 cd	22.4bc	10.2de	49.4b	35.6 cd
A. niger	9.1b	12.7d	14.7 cd	12.1c	14.5d	18.8d
Control	12.3a	18.9b	34.3bc	19.7b	47.8b	82.7a
P. oxalicum	0.0 cd	7.5d	10.6d	12.3c	14.1d	13.2e
Control	6.1bc	10.2 cd	27.1d	22.4a	44.0bc	65.9abc
S. rolfsii	3.9c	7.1d	8.3de	8.5e	9.8e	13.0e
Control	8.8b	13.5c	39.3b	14.8abc	28.3 cd	73.7bc
F. oxysporum	0.0 cd	3.4e	8.7de	8.6e	9.1e	15.4de
Control	5.4bc	21.9a	44.7a	15.9abc	38.7c	53.7c
Std deviation	4.6	5.5	13.1	4.8	17.2	28.4

Means with the same letter along a column are not significantly different (P=0.05).

Dania et al.

Dania et al. SpringerPlus 2014 3:711, doi:10.1186/2193-1801-3-711

Replicate tubers that were inoculated with *L. theobromae* and *R. solani* and treated with extracts differed significantly (P≤0.05) except for treatments that were bored and exposed. Rot initiation also differed significantly (P≤0.05) between treatments that were inoculated and exposed and those that were inoculated and covered. The extract was most effective in rot inhibition in tubers inoculated with *S. rolfsii* where rot was reduced to≤13.0% and least effective in suppressing *A. niger* rot (≤ 18.8%). *Lasiodiplodia theobromae* and *A.niger* were most virulent in control tubers without extract application with 78.2% and 82.75% rot respectively.

Quercus phillyraeoides extract effectively inhibited rot in all the replicate treatments of *R. solani*that were scratched, inoculated and exposed (Table 3), but could not totally inhibit the pathogen rot in treatments that were wounded, bored and exposed or covered.

However, the extract significantly (P≤0.05) reduced rot of all the other pathogens relative to the control treatments in the *in vivo* trials. The extract was least effective against rot development in tubers that were bored, inoculated and either exposed or covered.

Table 3: Rot inhibition in tubers inoculated with six virulent fungi 48 h before the application of *Q. phillyraeoides* extract stored for 6 months

Tuber rot (%)						
Fungus	Inoculated and exposed			Inoculated and covered		
	Sratch	Wound	Bore	Scratch	Wound	Bore
L. theobromae	9.9bc	14.0 cd	16.7bcd	13.8bc	17.1de	20.9de
Control	17.7a	29.8a	63.6a	19.5a	74.4a	69.8bc
R. solani	0.0e	11.5d	13.3d	10.5c	14.0cde	15.6e
Control	5.8 cd	16.5c	39.7abc	12.0bc	52.5 cd	45.2e
A. niger	7.5c	13.1 cd	19.6 cd	7.2 cd	20.6d	20.8de
Control	14.6b	25.6b	48.9bc	13.4bc	55.1c	78.3b
P. oxalicum	7.8c	10.7d	11.1de	8.3 cd	13.9cde	14.9e
Control	13.3b	22.3bc	32.4c	15.5b	48.2c	44.8 cd
S. rolfsii	4.7 cd	7.5e	11.3de	7.5 cd	17.3de	24.6d
Control	10.2bc	21.4bc	50.2b	19.8a	64.6b	81.7a
F. oxysporum	5.4 cd	11.5d	14.4d	10.5c	13.2cde	19.7de
Control	9.8bc	21.1c	47.6bc	16.6b	58.8bc	57.7abc
Std deviation	4.8	6.9	18.6	4.3	23.4	25.3

Means with the same letter along a column are not significantly different (P=0.05).

Dania et al.

Dania et al. SpringerPlus 2014 3:711, doi:10.1186/2193-1801-3-711

The extract was most effective in rot suppression in tubers inoculated with *P.oxalicum* where rot was reduced to≤14.9% and least effective in inhibiting *L. theobromae* (≤ 20.9% rot). *Sclerotium rolfsii* and *A. niger* were most virulent in control tubers with 81.7% and 78.3% rot respectively. There was significant difference in the control of rot development between the rot-causing fungi in tubers that were inoculated and exposed and those inoculated and covered (P≤0.05). There was no significant difference in the effect of *Q. phillyraeoides* on rot initiation in treatments that were wounded and exposed (P>0.05), except for tubers inoculated with *L. theobromae* and *A. niger* and*S.*

rolfsii which showed significant difference (P≤0.05). The efficacy of *Q. phillyraeoides* in the control of rot in tubers that were bored, inoculated and exposed differed significantly (P≤0.05). Similarly, there was a significant difference among all treatments that were either wounded or bored and exposed (P≤0.05). All treatments were significantly different from the control set up in all trials (P≤0.05).

Phytochemical Screening of Plant Extracts

Phytochemical analysis of *Oryza sativa* husk and *Quercus phillyraeoides* extract revealed the presence of various active constituents (Table 4). Both extracts were found to contain alkaloids, flavonoids, saponins, tannins, terpenoids, phenol and anthraquinone. However, Phlobatanins and ferulic acid were lacking in *Q. phillyraeoides* extract. Similarly, pyroligneous acid was absent in Oryza sativa husk extract.

Table 4: Phytochemical analysis of plant extracts

Phytochemical analysis of plant extracts		
Exract concentration mg/ml		
Phytochemical	O. sativa	Q. phillyraeoides
Alkaloids	2.16 ± 0.15	1.07 ± 0.03
Flavonoids	0.17 ± 0.20	0.187 ± 1.42
Saponins	3.30 ± 0.23	1.99 ± 0.02
Tannins	0.77 ± 0.11	0.17 ± 0.01
Phlobatanins	0.35 ± 0.01	0.00
Ferulic acid	5.69 ± 0.08	0.00
Terpenoids	0.71 ± 0.43	0.38 ± 0.05
Phenol	3.21 ± 0.04	2.59 ± 0.01
Anthraquinone	0.07 ± 0.01	0.45 ± 0.05
Pyroligneous acid	0.00	7.74 ± 0.12

Ferulic acid had the highest concentration of 5.69 mg/ml amongst the bioactive compounds contained in Oryza sativa husk extract, while pyroligneous acid had the highest concentration of 7.74 mg/ml in Quercus phillyraeoides extract.

Dania et al.

Dania et al. SpringerPlus 2014 3:711 doi:10.1186/2193-1801-3-711

DISCUSSION

Naturally, there are many plants for use in natural crop protection in order to ensure safe and healthy environment. These botanicals provide a viable alternative to synthetic pesticides because they are harmless to beneficial soil microorganisms such as symbiotic nitrogen-fixing bacteria, they are cheap and readily available, and they do not accumulate in the food chain. *Oryza sativa*extract used in this study completely inhibited the mycelial growth and conidia production in *A. niger, P. oxalicum*, and *F. oxysporum* at 3.5% w/v concentrations *in vitro* as revealed in Table 1. Comparatively, control plates showed distinct morphological features such as cluster of conidiospores borne in sterigmata and phaliades for *A. niger,* and *P. oxalicum*, respectively under microscopic examination. *Fusarium oxysporum* had spindle-shaped conidia. Similarly, they effectively inhibited the production of pycnidia in *L. theobromae* and the formation of`sclerotia in *R. solani* and *S. rolfsii*. These results agree with the findings of Abiala 2008 who reported the potential of *O. sativa* husk extract for the *in vitro* control of the black sigatoka pathogen, *Mycosphaerella fijiensis* Also, Dhaliwal et al. (1993) found extracts from several rice varieties to be effective in inhibiting *R. solani*.

The potential of plant extracts in thc control of plant pathogens is facilitated by the presence of active ingredients in the form of secondary metabolites. Ferulic acid and Pyroligneous acid are the active ingredients found in *O. sativa* husk and *Q. phillyraeoides* extracts respectively (Yoshida et al. 2000). These compounds are biodegradable, do not incite pathogen resistance and are not toxic to man and livestock. It must be stressed, however, that these active ingredients are not solely responsible for fungitoxicity of the extracts to the test pathogens in this study as revealed in Table 4. The existence of a synergy between the active ingredients and other secondary metabolites such as alkaloids and phenols in plant extracts enhances their potential in plant disease control. Similarly, alkaloids have been implicated in the control of phytopathogenic bacteria (Olugbade and Odebiyi 1992). The ability of the extracts to effectively inhibit the mycelial growth of all the rot-causing fungi *in vitro* is an indication of their broad-spectrum activity.

The success of botanicals in the control of plant pathogens can be determined by the medium of extraction. Some authors have reported that organic solvents such as ethanol extract active ingredients in plant extracts better than water and that they are more effective in controlling plant pathogens (Fuerhake 1984; Lale and Abdulrahaman 1998). However, organic solvents are chemicals and the implication of chemical control in our natural ecosystem cannot be over emphasized. Consequently, aqueous extracts were used in both the *in vitro* and *in vivo* trials in this study which are easy to prepare, cheap and are not toxic. The efficacy of plant extracts is also influenced by the part of the plant that is used for extraction. The active ingredients are often more concentrated in plant storage organs such as roots, tubers, fruits, rhizomes and seeds. Peluola (2005) reported that the concentration of azadirachtin active ingredient was comparatively higher in neem seeds than the leaves or bark.

The presence of secondary metabolites in *O. sativa* and *Q. phillyraeoides* extracts caused the inhibition of radial growth and spore germination of the test pathogens *in vitro*, which confirms the report of previous workers (Dhaliwal et al. 1993; Ejechi and Souzey 1999). The difference recorded in the efficacy of the extracts may be due to the solubility of their active ingredients in water.

The concentration of these extract metabolites is also an important determinant of their ability to inhibit pathogens. Some pathogens have the ability to biodegrade phenols of some botanicals; this reason is responsible for their inability to control such pathogens. *Oryza sativa* husk extract significantly reduced tuber rot *in vivo* at lower concentration as shown in Table 2. These results are consistent with the findings of (Abiala 2008) who reported lower inhibitory concentration ≤ 5% w/v for this extract in the *in vitro* control of *M. fijiensis*.

Injury to yam tubers in the form of wounds enhances the entry of fungal pathogens. This normally occurs during weeding, harvesting, through the feeding lesions of nematodes and insects or while in transit to the market. Rot development was influenced by the degree of injury done to the tubers in this study. Tables 2 and 3 showed that tubers that were bored and inoculated with test pathogens had higher rot incidence than those that were wounded or scratched and inoculated. This result agrees with the findings of (Ogaraku and Usman 2008) who reported a direct relationship between the degree of damage to yam tubers and rot incidence. However, disease incidence and development are primarily facilitated by the presence of a virulent pathogen, susceptible host, and favourable environment (Manjula et al. 2005). There was significant difference ($P \leq 0.05$) between the extracts in the control of rot between inoculated tubers that were covered and those that were exposed. They were more effective in controlling rot in tubers that were inoculated and exposed, while rot control was less effective among inoculated tubers that were covered. The higher rot values obtained in replicate tubers that were inoculated and covered could be attributed to the high relative humidity present in the covered treatments, which favours disease development. Conversely, the reduced rot recorded in inoculated and exposed tubers may be attributed to the low relative humidity which discourages fungal infection. Generally, fungal biodeterioration increases at relative humidity greater than 60%, and their virulence diminishes with decreasing relative humidity below this level. These results agree with the findings of Ikotun (1983a) who reported high relative humidity as an important factor in rot initiation of yam tubers.

Post-harvest biodeterioration of ware yam largely starts from the field, resulting from the injuries on yam tubers before storage. The invading fungal pathogens are subsequently carried from the field to the yam barn. In most cases, these infections are latent and only become noticed after incubation and the infected tubers start to rot. However, besides human activities, insects pests, rodents and nematodes that predispose yam tubers to rot, other factors such as sprouting, tuber respiration, temperature and the degree of resistance or susceptibility of yam genotypes to invading pathogens also influence yam storage. Cultural practices such as continuous cropping also encourage the persistence of yam pathogens and enhance their transition from the previous cropping season to the next season. Temperature influences yam storability. Yams store best at 28 ± 2°C in Nigeria. Although storage at lower temperaturs reduces microbial rotting, it causes chilling damage to yam tubers. Storage at temperatures higher than 30°C increases the incidence of xerophilic fungi like *Aspergillus niger* and *Penillium oxalicum*. Therefore, correct manipulation of temperature during storage can also prolong the shelf life of ware yam.

CONCLUSIONS

Tuber wound is the major factor that predisposes pre and post-harvest yam to fungal biodeterioration. The depth of injury to the yam tubers influenced the degree of rot by the test pathogens in this study. Treatments that were bored and inoculated with rot-causing fungi exhibited the highest degree of rot, followed by the wounded and then the scratched tubers. A vital inference and advice to peasant farmers, who are the target beneficiary of this research, therefore, is for them to try as much as possible to avoid or reduce injury to yam tubers during harvest to the barest minimum to prevent losses from yam pathogens. The promising potential of both *O. sativa* and *Q. phillyraeoides* extracts in rot inhibition in this study recommends their development as prospective biopesticide formulation for tuber rot control of white yam.

AUTHORS' CONTRIBUTIONS

The authors, VOD, OOF, MA and PLK carried out the research work on the bioassay of Oryza sativa husk and Quercus phillyraeoides extracts against yam pathogens, participated in the phytochemical analysis for metabolites and drafted the manuscript. All authors read and approved the final version of the manuscript.

ACKNOWLEDGEMENTS

The authors are grateful to International Institute of Tropical Agriculture (IITA), Ibadan, Nigeria for providing the requisite facilities to conduct this research.

REFERENCES

1. Abiala M (2008) Evaluation of plant extracts and microbial agents in the in vitro control of Mycosphaerella fijiensis. M.Sc Dissertation. University of Ibadan. p 95
2. Aboagyne-Nuamah FSK, Offei EW, Cornelius RD, Bancroft J (2005) Severity of spoilage storage rots of white yam (*Dioscorea rotundata* Poir). Ann Appl Biol 147(2):183-190

3. Aidoo AK (2007) Identification of yam tuber rot fungi from storage systems at the Kumasi Central Market. A dissertation submitted to Faculty of Agriculture. K.N.U.S.T. p 79
4. Amadioha AC (2000) Fungitoxic effects of some leaf extracts against *Rhizopus oryzae* causing tuber rot of potato. Arch Phytopathol 7:1-9
5. Amusa NA, Adegbite AA, Mohammed S, Baiyewu RA (2003) Yam diseases and its management in Nigeria. Afr J Biotechnol 12:497-502
6. Arua EO (1981) Yam ceremonies and the values of Ohafia culture. Africa 51:694-705
7. Babaleye T (2003) Raising the status of the yam, a major food crop in West Africa. ANB-BIA Suppl Issue/Ed 463:1-3
8. Barnett HL, Hunter BB (1998) Illustrated Genera of imperfect Fungi. Macmillan Inc, New York. p 218
9. Coursey DG (1983) Yams. Handbook of Tropical Foods. In: Chan HT (ed) International Review of Yams, New York: Dekker. pp 555-601
10. Dania VO (2012) Evaluation of some plant extracts and microbial agents for the control of post-harvest fungal rot of yam (Dioscorea spp.) Ph.D Thesis. University of Ibadan. p 219
11. Dhaliwal GS, Pathlak MD, Vega CR (1993) Effect of plant extracts of rice varieties on insect pests and predator complex of rice. J Crop Sci 1993(7):51-64
12. Ejechi BO, Souzey JA (1999) Inhibition of bio-deterioration of yam tuber (*Dioscorea rotundata* Poir) in storage with phenolic extracts of *Acalypha hispida* burm F. leaves. J Stored Prod Res 35:127-134
13. Enikuomehin OA, Ikotun T, Ekpo EJA (1998) Evaluation of ash from some tropical plants of Nigeria for the control of *Sclerotium rolfsii* Sacc. on wheat (*Triticum aestivum* L.). Mycopathol 142:81-87
14. Fuerhake KJ (1984) Effectiveness and selectivity of technical solvents for the extraction of neem seed components with insecticidal activity. In: Schumutterer H, Ascher KRS, Rembold GTZ (eds) Natural pesticides from the neem tree and other tropical plants, Echborn: Proceedings of second International Neem Conference. pp 103-114

15. Ghana Organic Agricultural Network (GOAN) (1999) The organic farmer. Quarterly newsletter of GOAN. 7
16. Hycenth N (2008) Effects of different plant extracts in the control of yam rot induced by *Rhizopus stolonifer* on stored yam (*Dioscorea* sp.) in Yola, Adamawa state, Nigeria. Agric J 3(5):382-387
17. Ikotun T (1983) Post-harvest microbial rot of yam in Nigeria. Fitopathologie Bras 1983(8):1-7
18. Ikotun T (1989) Disease of yam tubers. Int J Trop Plant 7:1-21
19. Lale NES, Abdulrahaman TA (1998) Evaluation of neem (*Azadirachta indica* A. Juss) seed oil obtained by different methods of neem powder fot the management of *Callosobruchus maculatus*(F) (Coleoptera: Bruchidae) in stored cowpea. J Stored Prod Res 35(2):135-143
20. Manjula K, Mwangi M, Bandyopadhyay R (2005) Potentials of some bacteria and fungi as biocontrol agents of cassava and yam under laboratory and green house conditions. Afr Crop Sci Proc 7:1395-1400
21. Noon PA (1978) Storage and market diseases of yams. Trop Sci 1978:17-81
22. Ogaraku AO, Usman HO (2008) Storage rot of some yams (*Dioscorea* spp.) in Keffi and Environs, Nasarawa State, Nigeria. Patnsuk J 4(2):22-27
23. Okorji EC (1992) Economics of yam production in Southern Nigeria. Beur Trop Landwirsch Vet Me 30:77-84
24. Olugbade TA, Odebiyi CO, Aladesanmi JA (1992) Antibacteria alkaloids in *Crinum jagus*. Pharmacognosy 304:303-307
25. Peluola CO (2005) Evaluation of botanicals and microbial agents in the control of some pathogens of cowpea. Ph.D dissertation. University of Ibadan. p 159
26. Prakash A, Rao J, C.R.R.I Cuttack (1994) Biodeterioration of paddy seed quality due to insects and mites and its control using botanicals. Annual report of ICAR. Adhoc Scheme (1993–1994). India. p 37
27. Purseglove JW (1985) Tropical Crops. Monocotyledons. Longman, London. pp 97-117

28. Sofowora A (1993) Medicinal Plants and Traditional Medicines in Africa 4: Chichester John Wiley & Sons, New York, pp 97-145

29. Titteh JP, Saakwa R (1994) Prospects and constraints of yam production in Ghana. In: Ofori CG, Halm PC (eds) Proceedings of the 9th symposium of the International Society for Tropical root Crops, Ghana: Accra. pp 20-26

30. Yoshida T, Hirojuki T, Eiji T, Hiroshi K (2000) Effect of components of pyroligneous acid on several plant pathogenic fungi. Jpn J Crop Sci 98:196-197

Citations

CHAPTER 1

Mohammed Kasso and Mundanthra Balakrishnan, "Ecological and Economic Importance of Bats (Order Chiroptera)," ISRN Biodiversity, vol. 2013, Article ID 187415, 9 pages, 2013. doi:10.1155/2013/187415.

CHAPTER 2

Masahiro (Mh.) Osakabe, Ryuji Uesugi, and Koichi Goka, "Evolutionary Aspects of Acaricide-Resistance Development in Spider Mites," Psyche, vol. 2009, Article ID 947439, 11 pages, 2009. doi:10.1155/2009/947439.

CHAPTER 3

Oleke, J.M., Manyong, V., Mignouna, D., Isinika, A., Mutabazi, K., Hanna, R., & Sabelis, M. (2013). Ex-ante economic analysis of biological control of coconut mite in Benin. AgBioForum, 16(2), 161-169.

CHAPTER 4

Hiroshi Abe, Takeshi Shimoda, Jun Ohnishi, Soichi Kugimiya, Mari Narusaka, Shigemi Seo, Yoshihiro Narusaka, Shinya Tsuda, and Masatomo Kobayashi, Jasmonate-dependent plant defense restricts thrips performance and preference, doi:10.1186/1471-2229-9-97.

CHAPTER 5

Juan Alfredo Morales-Ramos and Maria Guadalupe Rojas, "A Modular Cage System Design for Continuous Medium to Large Scale In Vivo Rearing of Predatory Mites (Acari: Phytoseiidae)," Psyche, vol. 2014, Article ID 596768, 8 pages, 2014. doi:10.1155/2014/596768.

CHAPTER 6

S. M. Adnan, M. M. Uddin, M. J. Alam, et al., "Management of Mango Hopper, Idioscopus clypealis, Using Chemical Insecticides and Neem Oil," The Scientific World Journal, vol. 2014, Article ID 709614, 5 pages, 2014. doi:10.1155/2014/709614.

CHAPTER 7

Asgar Ebadollahi, Jalal Jalali Sendi, Alireza Aliakbar, and Jabraeil Razmjou, "Chemical Composition and Acaricidal Effects of Essential Oils of Foeniculum vulgare Mill. (Apiales: Apiaceae) and Lavandula angustifoliaMiller (Lamiales: Lamiaceae) against Tetranychus urticae Koch (Acari: Tetranychidae)," Psyche, vol. 2014, Article ID 424078, 6 pages, 2014, doi:10.1155/2014/424078.

CHAPTER 8

R Scott Cornman, Michael C Schatz, J Spencer Johnston, Yan-Ping Chen, Jeff Pettis, Greg Hunt, Lanie Bourgeois, Chris Elsik, Denis Anderson, Christina M Grozinger, and Jay D Evans, Genomic survey of the ectoparasitic mite Varroa destructor a major pest of the honey bee Apis mellifera, doi:10.1186/1471-2164-11-602.

CHAPTER 9

Wycliffe Wanzala, Ahmed Hassanali, Wolfgang Richard Mukabana, and Willem Takken, "Repellent Activities of Essential Oils of Some Plants Used Traditionally to Control the Brown Ear Tick, Rhipicephalus appendiculatus,"Journal of Parasitology Research, vol. 2014, Article ID 434506, 10 pages, 2014. doi:10.1155/2014/434506.

CHAPTER 10

Victor Ohileobo Dania, Olubunmi Omowunmi Fadina, Maria Ayodele, and P Lava Kumar, Efficacy of Oryza Sativa Husk and Quercus Phillyraeoidesextracts for The in Vitro and in Vivo Control of Fungal Rot Disease of White Yam (Dioscorea Rotundata Poir), doi:10.1186/2193-1801-3-711.

Index